Eckart Ehlers
Thomas Krafft
Earth System Science in the Anthropocene

Eckart Ehlers
Thomas Krafft
(Editors)

Earth System Science in the Anthropocene

In Collaboration with C. Moss (Linguistic editing)

With 37 Figures and 10 Tables

 Springer

Editors

Professor Dr. Eckart Ehlers
Department of Geography
University of Bonn
Meckenheimer Allee 166
53115 Bonn
Germany
E-mail: ehlers@giub.uni-bonn.de

Dr. Thomas Krafft
German National Committee
on Global Change Research
Scientific Secretariat
Luisenstr. 37
80333 München
Germany
E-mail: t.krafft@iggf.geo.uni-muenchen.de

Library of Congress Control Number: 2005927790

ISBN-10 3-540-26588-0 Springer Berlin Heidelberg New York
ISBN-13 978-3-540-26588-7 Springer Berlin Heidelberg New York

Springer is a part of Springer Science+Business Media
Springeronline.com
© Springer-Verlag Berlin Heidelberg 2006
Printed in The Netherlands

Cover design: E. Kirchner, Heidelberg
Typesetting: SPI Publisher Services, Pondicherry, India
Production: Almas Schimmel
Printing: Krips bv, Meppel
Binding: Stürtz AG, Würzburg

Printed on acid-free paper 32/3141/as 5 4 3 2 1 0

Preface

This volume intends to document some aspects of the ongoing debate on crucial issues of Earth System Science in the Anthropocene. To the best of the editors' knowledge it is still one of only those few publications that provide different perspectives on the concept of the "Anthropocene". A group of international and German experts – many of the latter members of the German National Committee on Global Change Research (NKGCF) – have contributed their views on current or emerging issues in the context of Global Change and Earth System Science, Environmental Change and Human Security, Interfaces between Nature and Society and on Scientific Challenges for Anthropocenic Research in the 21st Century. The volume provides an overview and update on the new Joint Projects of the Earth System Science Partnership of the four international Global Change Research Programmes and on some German contributions to them. As a sequel to the earlier publication "Understanding the Earth System" this book aims at stimulating the necessary scientific discussion on the future development of Global Change Research and the better integration of and cooperation between natural and social sciences.

The editors want to thank the authors of this volume for their contributions and all those who have made this book possible. They are especially grateful to the authors for their patience in connection with the sometimes painfully slow editing process of this volume. While every effort has been made on the part of the editors to ensure consistency in terminology, methods of quotation and other formalities in connection with the publication process, they surely did not manage to escape certain inconsistencies in the final version of the book.

Financial assistance for this publication has been kindly provided by the German Ministry for Education and Research (BMBF). Editors and authors are grateful for the Ministry's continuous commitment and support to Global Change Research.

Bonn, München, June 2005
Eckart Ehlers
Thomas Krafft

Contents

List of Contributors

Thomas Berger
Center for Development Research
University of Bonn
Walter-Flex-Str. 3
53113 Bonn, Germany
e-mail: t.berger@uni-bonn.de

Hans-Georg Bohle
Department of Geography
University of Bonn
Meckenheimer Allee 166
53115 Bonn, Germany
e-mail: bohle@giub.uni-bonn.de

Mike Brklacich
Global Environmental Change and
Human Security Project and Department
of Geography and Environmental
Studies,
Carleton University
1125 Colonel By Drive
Ottawa, Ontario, KIS 5B6
Canada
e-mail: michael_brklacich@carleton.ca

Paul J. Crutzen
Max-Planck-Institute for Chemistry
Joh.-Joachim-Becher-Weg 27
55128 Mainz, Germany
e-mail: air@mpch-mainz.mpg.de

Eckart Ehlers
Department of Geography
University of Bonn
Meckenheimer Allee 166
53115 Bonn, Germany
e-mail: ehlers@giub.uni-bonn.de

Nick van de Giesen
Center for Development Research
University of Bonn
Walter-Flex-Str. 3
53113 Bonn, Germany
e-mail: nick@uni-bonn.de

Johann G. Goldammer
Max-Planck-Institute for Chemistry
Joh.-Joachim-Becher-Weg 27
55128 Mainz, Germany
e-mail: johann.goldammer@.
fire uni-freiburg.de

Hartmut Graβl
Meteorological Institute,
University of Hamburg and
Max Planck Institute for Meteorology
Bundesstr. 53
20146 Hamburg, Germany
e-mail: grassl@dkrz.de

John Ingram
GECAFS International Project Office,
NERC-Centre for Ecology and
Hydrology
Wallingford, OX 10 8BB, United
Kingdom
e-mail: jsii@ceh.ac.uk

Maria Iskandarani
Center for Development Research
University of Bonn
Walter-Flex-Str. 3
53113 Bonn, Germany
e-mail: m.iskandarani@uni-bonn.de

Jill Jäger
Sustainable Europe Research Institute
(SERI)
Garnisongasse 7/27
1090 Vienna, Austria
e-mail: fuj.jaeger@nextra.at

Gernot Klepper
Kiel Institute for World Economics
Düsternbrooker Weg 120
24105 Kiel, Germany
e-mail: gklepper@ifw-kiel.de

Thomas Krafft
German National Committee on Global
Change Research,
Scientific Secretariat
Department of Earth and Environmental
Sciences, LMU Munich
Luisenstr. 37
80333 Munich, Germany
e-mail : nkgcf@iggf.geo.
uni-muenchen.de

Rik Leemans
Environmental Systems Analysis Group,
Wageningen University
P.O. Box 9101, 6700 HB Wageningen,
The Netherlands
e-mail: rik.leemans@wur.nl

Karin Lochte
Institute for Marine Research at the
University of Kiel
Düsternbrooker Weg 20
24105 Kiel, Germany
e-mail: klochte@ifm-geomar.de

Franziska Matthies
Tyndall Centre (HQ),
School of Environmental Sciences,
University of East Anglia
NR4 7TJ Norwich, United Kingdom
e-mail: f.matthies@uea.ac.uk

Wolfram Mauser
German National Committee on Global
Change Research
Department of Earth and Environmental
Sciences, Section Geography, LMU
Munich
Luisenstr. 37
80333 Munich, Germany
e-mail : w.mauser@iggf.geo.
uni-muenchen.de

James K. Mitchell
Department of Geography
Rutgers University
Piscataway, NJ 08854-4085, USA
e-mail: jmitchel@rci.rutgers.edu

Horst J. Neugebauer
Geodynamics, University of Bonn
Nussallee 8
53115 Bonn, Germany
e-mail: neugb@geo.uni-bonn.de

Soojin Park
Center for Development Research
University of Bonn
Walter-Flex-Str. 3
53113 Bonn, Germany
e-mail: spark@uni-bonn.de

Erich J. Plate
Emeritus Professor of Civil Engineering
University of Karlsruhe (TH), Germany
c/o Institut für Wasserwirtschaft
u. Kulturtechnik
Kaiserstr. 12
76128 Karlsruhe, Germany
e-mail: erich.plate@bau-verm.
uni-karlsruhe.de

Rainer Sauerborn
Department of Tropical Hygiene
and Public Health,
University of Heidelberg
Im Neuenheimer Feld 324
69120 Heidelberg, Germany
e-mail: rainer.sauerborn@urz.
uni-heidelberg.de

Paul Vlek
Center for Development Research
University of Bonn
Walter-Flex-Str. 3
53113 Bonn, Germany
e-mail: p.vlek@uni-bonn.de

Alfred Voβ
Institute of Energy Economics and
the Rational Use of Energy,
University of Stuttgart
Heβbrühlstr. 49a
70550 Stuttgart, Germany
e-mail: av@ier.uni-stuttgart.de

Wolfgang-Peter Zingel
South Asia Institute,
University of Heidelberg,
Department of International Economics
Im Neuenheimer Feld 330
69120 Heidelberg, Germany
h93@ix.urz.uni-heidelberg.de

Acronyms

AIDS	Acquired Immunodeficiency Syndrome
ASEAN	Association of South East Asian Nations
BIBEX	Biomass Burning Experiment
BMBF	German Federal Ministry of Education and Research
CBD	Convention on Biological Diversity
cCASHh	Climate Change and Adaption Strategies for Human Health in Europe
CCN	Cloud Condensation Nuclei
CDM	Clean Development Mechanism
CER	Cumulative Energy Requirements
CGIAR	Consultative Group on International Agricultural Research
CIMMYT	International Maize and Wheat Improvement Centre
COP1; COP3; COP7	First; Third; Seventh Session of the Conference of the Parties to the UNFCCC
CSF	Common Sampling Frame
DALY	Disability Adjusted Life Years
DFG	Deutsche Forschungsgemeinschaft
DIVERSITAS	International Programme of Biodiversity Science
DPSIR	Driver-Pressure-State-Impact-Response
DSS	Decision Support System
ECO	Economic Cooperation Organisation
ENSO	El Nino-Southern Oscillation
EPP	Energy Payback Period
ESSP	Earth System Science Partnership
FAO	Food and Agriculture Organisation of the United Nations
FAO-ECE	Food and Agriculture Organisation Economic Commission for Europe
FCCC	*see UNFCCC*
GC	Global Change
GCMs	General Circulation Models
GCP	Global Carbon Project
GDP	Gross Domestic Product
GEC	Global Environmental Change
GECAFS	Global Environmental Change and Food Systems
GFMC	Global Fire Monitoring Center
GHG	Green House Gases
GIS	Geographic Information System

GLOWA	Global Change in the Hydrological Cycle
GLSS	Ghana Living Standards Survey
GOFC	Global Observation of Forest Cover
GOLD	Global Observation of Land Dynamics
GPP	Gross Primary Product
GTOS	Global Terrestrial Observing System
HDI	Human Development Index
HELP	Hydrology for the Environment, Life and Policy
IAM	Integrated Assessment Models
ICRP	International Commission on Radiological Protection
ICSU	International Council for Science
ICT	Information and Communication Technologies
IDNDR	International Decade for Natural Disaster Reduction
IFFN	International Forest Fire News
IGAC	International Global Atmospheric Chemistry
IGBP	International Geosphere-Biosphere Programme
IGP	Indo-Gangetic Plain
IHDP	International Human Dimensions Programme on Global Environmental Change
INDEPTH	International Network of field sites with continuous Demographic Evaluation of Populations and Their Health in developing countries
IPAT	Impacts =Population * Affluence * Technology
IPCC	Intergovernmental Panel on Climate Change
IPCC–TAR	IPCC Third Assessment Report
ISDR	International Strategy for Disaster Reduction
ISSC	International Social Science Council
LCA	Life Cycle Assessment
LDC	Least Developed Countries
LULUCF	Land-Use, Land-Use Change, and Forestry
MA	Millennium Ecosystem Assessment
MARA	Mapping Malaria Risk in Africa
MDC	More Developed Countries
NAFTA	North American Free Trade Agreement
NGO	Non-Governmental Organisations
NKGCF	German National Committee on Global Change Research
NRC	United States National Research Council
OECD	Organisation for Economic Co-operation and Development
PCA	Principal Component Analysis
PFDS	Public Food Distribution System
R&D	Research and Development
S&T	Science and Technology
SAARC	South Asian Association of Regional Cooperation
SAFARI	Southern Africa Fire-Atmosphere Research Initiative

SARS	Severe Acute Respiratory Syndrome
SRES	IPCC Special Report on Emissions Scenarios
STARE	Southern Tropical Atlantic Regional Experiment
SWCC	Second World Climate Conference
UNCCD	United Nations Convention to Combat Desertification
UNEP	United Nations Environmental Programme
UNESCO	United Nations Educational, Scientific and Cultural Organization
UNFCCC	United Nations Framework Convention on Climate Change
UNFF	United Nations Forum of Forests
USAID	United States Agency for International Development
WAPP	West Africa Power Pool
WBGU	German Advisory Council on Global Change
WCRP	World Climate Research Programme
WHO	World Health Organisation
WMO	World Meteorological Organisation
WPI	Water Poverty Index
WSSD	World Summit for Sustainable Development
YLD	Years Lived with Disability
YLL	Years of Life Lost

Part 1

Introduction: The Anthropocene

1.0 Global Change Research in the Anthropocene: Introductory Remarks

Wolfram Mauser

Chair German National Committee on Global Change Research (NKGCF),
Department of Earth and Environmental Sciences, Section Geography, LMU Munich,
Luisenstr. 37, 80333 Munich

Global Change summarises the growing interference of human beings with the Earth's metabolism and its relation to the natural variability of the Earth System. The human interference in the Earth System strongly increased during the last century and has now reached a level which is of the order of magnitude as many natural processes on Earth. The term Anthropocene has been suggested to mark an era in which the human impact on the Earth System has become a recognisable force. Presently there are no signs for a deceleration or reversal of this development. Coping with the consequences of Global Change therefore becomes a challenge of prior unknown dimension for human societies. It goes far beyond the analysis of changes in the global climate system and rather comprises the Earth System with its physical, bio-geochemical and societal processes as a whole. Humans at the same time cause, are affected by, and alter change. The causes, mechanisms and effects of interactions between humans and the other components of the Earth System are therefore at the very centre of Global Change research.

Global Change research is especially devoted to the understanding of the evolution and the impacts of the processes that trigger and drive change and to the derivation of possible alternatives for actions to deal with change. Improved observation systems for the components of the Earth System, the analysis of natural variability and complex interactions as well as the development of prognostic abilities to foresee changes and their consequences are essential to identify options for favourable counteractions and to develop strategies for adaptation to change as well as to mitigate causes.

There is no alternative to sustainable development as the long term strategy for coping with the consequences of Global Change. Although sustainability sometimes seems to be rather vague in its content and meaning, common understanding has been achieved that it is based on an intelligent balance of current needs and the interests of future generations. Ecology and the environment are therefore equally important as economic and social aspects. Despite the clear theoretical evidence that the development of humankind and its interaction with other components of the Earth System have to follow a sustainable pathway, we have underestimated the intellectual complexity related to the questions of what exactly sustainable development is and how suitable pathways to sustainability can be identified. The necessary scientific progress can only be achieved if we openly address these questions and if we cross disciplinary boundaries wherever and whenever necessary developing a common understanding beyond disciplinary terminologies.

The effects of Global Change are manifested in strong geographic differentiations. It is on the regional/national level that decisions and actions towards sustainable development will most likely be implemented. Global Change Research must extend the common efforts to bridge the gap between the global perspective on change and the analysis of regional impacts and management options. Despite the many unsolved questions we are still facing, Global Change Research under the umbrella of ICSU's and ISSC's four international Global Change research programmes (WCRP, IGBP, DIVERSITAS and IHDP) has achieved significant scientific progress. The four programmes have facilitated a new and changed view on Earth. Their increasing cooperation under the common roof of the Earth System Science Partnership will inevitably lead to a better understanding of the complex interactions within the Earth System. But still, there remains a certain scientific perplexity when studying the complex issues of regional impacts of Global Change and suitable alternatives for actions to be taken on the way towards sustainability. In the long run the success of Global Change research will be judged by the ability to overcome this perplexity and by the amount of applicable knowledge and advice it has provided on the critical issue of the interaction of human beings with the other components of the Earth System.

"Earth System Science in the Anthropocene" is published in collaboration with the German National Committee on Global Change Research (NKGCF) that advises the Deutsche Forschungsgemeinschaft (DFG) and the Federal Ministry of Education and Research (BMBF) on research related issues of Global Change. The book contains views from different fields of Global Change Research on the scientific challenges in the Anthropocene. It also reflects the sometimes controversial but always stimulating deliberations within the National Committee on the future development of Global Change Research. This volume intends to reach out to the scientific community and interested public to encourage participation in the necessary scientific debate.

1.1 Managing Global Change: Earth System Science in the Anthropocene

Eckart Ehlers[1], Thomas Krafft[2]

[1]Department of Geography, University of Bonn, Meckenheimer Allee 166, 53115 Bonn
[2]German National Committee on Global Change Research, Scientific Secretariat, Department of Earth and Environmental Sciences, LMU Munich, Luisenstr. 37, 80333 Munich

It is only a few years since terms like "Anthropozoikum" and "Anthropocene" have come into existence – or, at least, have attracted the attention of a broader academic public[1]. Embedded into the discussions around climate change and/or the many facets of Global Change, "Anthropozoikum" and "Anthropocene" are considered to be scientific terminologies that may be suited to stand for the beginning of a potentially new geological era: an era dominated by the increasingly stronger and obviously lasting imprint of mankind on nature. This human fingerprint and its impacts on the climate system and on natural human environments may well have the dimensions and consequences of a geological force.

In his contribution to this volume, Paul J. Crutzen suggests that the beginning of the Industrial Revolution may also mark the beginning of the Anthropocene characterised as an "in many ways human-dominated geological epoch supplementing the Holocene" or more precisely, that the accelerating human activities like land use changes, deforestation, and fossil fuel burning since the later 18th century have driven multiple interacting effects that alter the earth's environment on an unprecedented scale.[2] It coincides with scientific evidence of growing concentrations of carbon dioxide, methane and other greenhouse gases in the atmosphere as well as in terrestrial or marine deposits.

Crutzen provides a persuasive list of human activities that influence our climate system and our natural environments upon which human societies and cultures depend. It has repeatedly been stated that these influences or changes create unprecedented challenges that demand new strategies to generate scientific

---

[1] References to these two terms can be found in Markl (1986: Anthropozoikum) and especially in Crutzen (2000: Anthropocene) where there is also a very short history of the ideas about human interferences with the Earth System and its role as a "telluric force".

[2] For more detailed analyses of the environmental impacts of fossil energies cf. Fischer-Kowalski and others 1997, Sieferle 1982, 1997 and 2001.

knowledge. Global Change and its impact on the food supply, water availability, decent human living conditions such as health, sanitation, education, housing etc. for a rapidly growing world population are probably the most demanding challenges of the 21st century. In spite of the extreme complexities of interactions and interdependencies of the Earth System to the increasing anthropogenic forcing, sustainable solutions must be found to reconcile nature and society. This, however, is easier said than done.

The earliest serious attempts to identify human causes of environmental changes and their lasting imprint on nature go back to the mid 19th century; in a less scientific way even to antiquity. While Greek and Roman authors give vivid reports about man's devastating impacts on their Mediterranean "Arcadia"[3], it is probably G.P. Marsh who, in 1864, delivered the first comprehensive analysis of a predominantly human-driven transformation of the earth surface. His "Man and Nature; Or, The Earth as modified by Human Action" is an almost visionary anticipation of the late 20th-century research agenda. In the preface to his book Marsh (1864, p. 3) writes:

> "The object of the present volume is: to indicate the character and, approximately, the extent of the changes produced by human action in the physical conditions of the globe we inhabit; to point out the dangers of imprudence and the necessity of caution in all operations which, on a large scale, interfere with the spontaneous arrangements of the organic or the inorganic world; to suggest the possibility and the importance of the restoration of disturbed harmonies and the material improvement of waste and exhausted regions; and, incidentally, to illustrate the doctrine, that man is, in both kind and degree, a power of a higher order than any of the other forms of animated life, which, like him, are nourished at the table of bounteous nature".

And almost all problems of today are already covered in Marsh's analysis, and probably for the first time in a highly comprehensive and systematic matter: the transfer, modification and extirpation of plant and animal life (ch. 2), the woods (ch. 3) and the waters (ch. 4) and even discussions about good and bad governance, uncertainties of meteorological knowledge or the restoration of disturbed harmonies.

While Marsh's stimulating book fell more or less into oblivion for almost a century, two post World War II publications revived the discussions about the human impact on the natural environment: "Man's Role in Changing the Face of the Earth", edited by W.L. Thomas Jr. in 1956, and "The Earth as Transformed by Human Action. Global and Regional Changes in the Biosphere over the Past 300 Years", edited by B.L. Turner II as chief-editor in 1990. It is probably not unfair to say that especially the latter publication has been instrumental and – to a certain aspect – also the turning-point towards a more balanced view of the driving forces behind

---

[3] Plato's famous description of the deterioration of Athen's natural environment in his "Kritias" serve as an example. Other examples are Hippocrates or Roman authors like Strabo and Plinius.

the obvious changes in the Earth System. Although a great number of other studies on human-driven environmental transformation processes have been published before this book (authors and editors make reference to them!), and not only in America, none of those publications have gained the same attention and impact beyond disciplinary boundaries than "The Earth As Transformed by Human Action". The balanced consideration of nature as well as society as potential drivers of climate and environmental changes as well as the highly commendable attempt to reconcile both ends of the transformation processes is probably one of the main reasons for the wide recognition and impact of the book.

Since 1990 a great number of studies have appeared which increasingly stress the human influence in many critical aspects of Global Change. It would go far beyond the intentions and potentials of these introductory remarks to mention all of them.[4] The hitherto ultimate acknowledgement of the human role in the transformation not only of the earth's surface but also of the earth and climate system is the declaration of the First Open Science Conference on "Challenges of a Changing Earth" held in Amsterdam in July 2001. The final and generally approved document of this conference acknowledges among other findings, the following facts:

- "Human activities are significantly influencing Earth's environment in many ways in addition to greenhouse gas emissions and climate change. Anthropogenic changes to Earth's land surface, oceans, coasts and atmosphere and to biological diversity, the water cycle and biogeochemical cycles are clearly identifiable beyond natural variability. They are equal to some of the great forces of nature in their extent and impact. Many are accelerating. Global change is real and is happening now."
- "Global change cannot be understood in terms of a simple cause-effect paradigm. Human-driven changes cause multiple effects that cascade through the Earth System in complex ways. These effects interact with each other and with local- and regional-scale changes in multidimensional patterns that are difficult to understand and even more difficult to predict. Surprises abound."

---

[4] An impressive document is the publication of the NRC (1999) "Our Common Journey. A Transition Toward Sustainability" in which needs and necessities of interactive/transdisciplinary research designs are discussed. See also NRC/NAP (1999). An equally stimulating study is the four-volume edition of "Human Choice and Climate Change", ed. by St. Rayner – E.L. Malone (1998), to mention only two.

The Third Assessment Report of the IPCC itself is a vivid proof of the growing acceptance of a decisively human imprint on nature and environment. And the impressive 5-volume "Encyclopaedia of Global Environmental Change" recognizes the role of the human factor to such a degree that its fifth volume is entirely devoted to "Social and Economic Dimensions of Global Environmental Change". And also volume 4 (Responding to Global Environmental Change) is very strongly influenced by socio-economic perspectives, thus counterbalancing the first two volumes which deal with the physical, chemical and biological dimensions of global environmental changes.

- "Earth System dynamics are characterised by critical thresholds and abrupt changes. Human activities could inadvertently trigger such changes with severe consequences for Earth's environment and inhabitants. The Earth System has operated in different states over the last half million years, with abrupt transitions (a decade or less) sometimes occurring between them. Human activities have the potential to switch the Earth System to alternative modes of operation that may prove irreversible and less hospitable to humans and other life. The probability of a human-driven abrupt change in Earth's environment has yet to be quantified but is not negligible."

And, logically, it concludes:

- "A new system of global environmental science is required. This is beginning to evolve from complementary approaches of the international global change research programmes and needs strengthening and further development. It will draw strongly on the existing and expanding disciplinary base of global change science; integrate across disciplines, environment and development issues and the natural and social sciences; collaborate across national boundaries on the basis of shared and secure infrastructure; intensify efforts to enable the full involvement of developing country scientists; and employ the complementary strengths of nations and regions to build an efficient international system of global environmental science." (Steffen W et al. (eds) 2002, pp 207-8)

The different attempts at conceptualising a "Sustainability Science" (cf. J. Jaeger's article in this volume) and the creation of an "Earth System Science Partnership" (ESSP) (Figure 1.1.1) may be considered as important outcomes of the Amsterdam Conference and can be seen as a major step towards a new bridging of the gap between the natural and social sciences. To meet the need for applicable results, disciplines have started to change and mutate – and continue to do so! Global Change research has stimulated "import-export exchange" of concepts and methodologies with changing scientific coalition partners and changing networks. The growing number of hybrid departments in the worldwide university system proves that academia is also adapting to the new challenges and demands of problem and solution oriented complexities of research. The complexities of the Earth System dynamics (and also those of sustainability science) can only be tackled through scientific cooperation not only beyond disciplinary boundaries, but also <u>across</u> disciplines and academic fields. Whether we call these new coalitions "goal-oriented multi-disciplinarity", "problem-oriented interdisciplinarity" or "self-reflexive transdisciplinarity"[5]: such reflections are almost superfluous in view of the urgencies of problems to be solved.

It would be tempting to go into a closer discussion of origins, causes and effects of the deplorable loss of the "unity of knowledge" – to paraphrase the title of a

---

[5] For a discussion of these and other concepts from a social science perspective cf. E. Becker – UNESCO, Paris 1997.

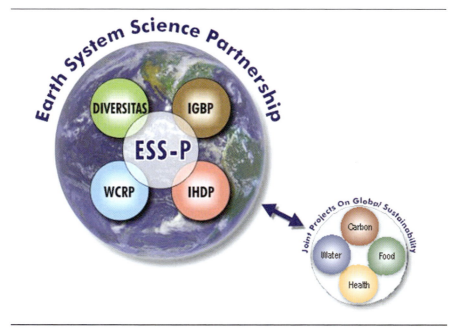

Fig. 1.1.1 The structure of the Earth System Science Partnership (ESSP)

book by E. O. Wilson (1998)[6]. The differentiation of science and its disciplinary categorisation into "mémoire" (history), "raison" (philosophy) and "imagination" (poetry) by the French encyclopaedists have greatly contributed to an unprecedented development of science, the humanities and the arts from 1780 onwards[7]. It included also the development and promotion of disciplines and disciplinary thinking as a precondition for the huge advances of science and technology in the 19th and 20th centuries. Today, we have to admit that the extremely specialised results of basic science and research are only partly and to a limited extent suited for the solution of all those challenges resulting from Global Change. The "discovery" of the human dimensions as one of the main driving forces of Global Change has added to the complexity of the problems at stake. The scientific term "Anthropocene" is an almost self-evident proof of this new complexity of interactions and interrelationships between the natural and the social systems.

It is, therefore, encouraging to see that also from a philosophical-theoretical point of view pragmatism begins to prevail. Very recently, P. Janich (2002) has

---

[6] For an interesting interpretation of the increasingly dramatic divergence of natural and social sciences see Bruno Latour "We have never been modern" (1993).

[7] For a comprehensive analysis of encyclopaedic developments in France (and its predecessors in England) see e.g. D.N. Livingstone, 1995. The original discourse can be found in Denis Diderot: "Discours Préliminaire" in "Encyclopédie, ou Dictionaire raisonné des sciences, des arts et des métiers, par une société de gens de lettres". Paris (Gide) 1751.

reflected on "Mensch und Natur" under the specific perspective of a "Revision eines Verhältnisses im Blick auf die Wissenschaften". His conclusions in regard to the commensurability of naturalism and culturalism show remarkable parallels with those more pragmatic approaches as expressed in the US National Research Council (NRC) publication "Our Common Journey". One of Janich's final conclusions is that of a division of labour through a methodological order (which is discussed in great detail in the paper!) – an approach not so far removed from NRC's proposal to integrate knowledge and action in the pursuit of a pathway towards sustainability (NRC, p. 10):

> "Because the pathway to sustainability cannot be charted in advance, it will have to be navigated through trial and error and conscious experimentation. The urgent need is to design strategies and institutions that can better integrate incomplete knowledge with experimental action into programs of adaptive management and social learning."

The scientific knowledge needed for identifying suitable pathways towards sustainability must derive from a broad scope of both disciplinary and interdisciplinary research, across the natural, social, economic, and engineering science domains. Earth System science has to provide place-based information by analysing global and regional processes of Global Change and by translating the research findings into policy relevant results.

As only one case in point here may serve the important issue of Global Change and urban health. Despite a long history of urban sanitary reform and healthful-city movements, inhabitants of rapidly growing urban agglomerations in the developing world are increasingly confronted with severe environmental health risks. The rapid urbanisation process experienced by the majority of developing countries during the last few decades has resulted in fundamental changes to the environment as well as to the social structure and is affected by and at the same time contributes to Global Change. While these changes affect all rapidly growing urban areas in one way or the other, in most of the megacities that have grown to unprecedented size the pace of urbanisation has by far exceeded the growth of the necessary infrastructure and services. As a result, an ever-increasing number of urban dwellers are left without access to basic amenities and face appalling living conditions. Additionally, social inequalities lead to subsequent and significant intra-urban health inequalities. Land use changes resulting from urban expansion often create changes in environmental conditions and in habitat for a number of species, which can trigger the outbreak of diseases. Overcrowding in urban agglomerations provides an easy pathway for the spread of communicable diseases. Large-scale migration to urban areas and integration into a global market where borders are frequently crossed and large distances are easily travelled by a growing number of people, allow the fast movement of infected individuals into previously unexposed populations. The recent outbreak of the Severe Acute Respiratory Syndrome (SARS), and its rapid international spread highlights the vulnerability of large urban agglomerations to new, emerging diseases, in a globalised world (Krafft T, Wolf T, Aggarwal S 2003).

The interactions and accelerating effects of both Global Change and rapid urbanisation create conditions in which the city can become "a dynamo driving infection" (Horton R 1996, p 135). Urban agglomerations and urban health care systems have also shown increasing vulnerability to the impacts of extreme weather effects. In a persuasive editorial for The Lancet, R. Horton has called for new integrative research initiatives to provide sustainable solutions for these problems: "the need for careful collaboration between medical geographers, epidemiologists, microbiologists, and infectious disease physicians is an urgent priority. [...] Unparalleled interdisciplinary cooperation and long-term governmental perspective are needed if we are to prevent the grand metropolis from becoming a sick and diseased necropolis" (Horton 1996, p 135).

The international Global Change research community is taking on these challenges. New integrative initiatives by both the Earth System Science Partnership (new joint initiatives on Global Water Systems, Food Systems, Carbon Regimes and Health) and the four major International Global Change Research Programmes (e.g., new urbanisation core project of IHDP) are aimed at filling some of the critical gaps. These initiatives are part of a major process of reorganising, refocusing and strengthening the international efforts on Global Change research based on the aforementioned Amsterdam principles.

The Anthropocene demands an Earth System Science, which understands humankind as integrated part of the Earth System. Emerging is an Earth System Science in a framework that encompasses all aspects of Global Change that may be well suited to provide essential answers for identifying suitable pathways towards sustainability.

References

Becker E et al. (1997) Sustainability: A Cross-Disciplinary Concept for Social Transformations. MOST Policy Papers No. 6. UNESCO, Institut für sozial-ökologische Forschung ISOE. Paris (F)

Clark W, Kates RW and members of the Board on Sustainable Development, NRC (1999) Our Common Journey: A Transition Toward Sustainability. Washington, D.C.

Crutzen PJ, Stoermer EF (2000) The "Anthropocene". IGBP Newsletter, 41, 17-18

Diderot D, Alembert JLR (1751) Encyclopedie, ou Dictionnaire raisonnè des science, des arts et des mètiers, par une societè de gens de lettres. Paris

Fischer-Kowalski M, Haberl H, Hüttler W et al. (1997) Gesellschaftlicher Stoffwechsel und Kolonisierung von Natur. Ein Versuch in Sozialer Ökologie. Amsterdam: Gordon & Breach Fakultas

Horton R (1996) The infected metropolis. The Lancet 347:134-135

Janich P(2002) Mensch und Natur. Zur Revision eines Verhältnisses im Blick auf die Wissenschaften, In: Sitzungsberichte der Wissenschaftlichen Gesellschaft an der Johann Wolfgang Goethe-Universität Frankfurt am Main, Bd. XL, 2, (2002) 26 S

Krafft T, Wolf T, Aggarwal S (2003) A new urban penalty? Environmental and health risks in Delhi. Petermanns Geographische Mitteilungen 147(4):20-27

Latour B (1993) We Have Never Been Modern. Cambridge, Mass, Harvard University Press

Livingstone DN (1995) The Spaces of Knowledge: Contributions Towards a Historical Geography of Science. Environment Planning D 13, 1995, 5-34

Markl, H. (1986): Natur als Kulturaufgabe. Deutsche Verlags-Anstalt/Stuttgart: 391 S.

Marsh GP (1864) Man and Nature; or Physical Geography as Modified by Human Nature. London

Rayner S, Malone EL (eds) (1998) Human Choice and Climate Change. Vol. I-IV. Columbus/Ohio

Sieferle RP (1982) Der unterirdische Wald. Energiekrise und Industrielle Revolution. -München

Sieferle RP (1997) Rückblick auf die Natur. Eine Geschichte des Menschen und seiner Umwelt. -München

Sieferle, R.P. (2001) The Subterranean Forest: Energy Systems and the Industrial Revolution. Cambridge

Steffen W et al. (eds) (2002) Challenges of a Changing Earth. Proceedings of the Global Change Open Science Conference, Amsterdam, The Netherlands, 10-13 July 2001. Berlin Heidelberg New York

Thomas WL (1956) Man's Role in Changing the Face of the Earth. Chicago

Turner BL et al. (eds) (1990) The Earth as Transformed by Human Action. Cambridge

Wilson EO (1998) The Unity of Knowledge. New York

1.2   The "Anthropocene"

Paul J. Crutzen

Max-Planck-Institute for Chemistry, Joh.-Joachim-Becher-Weg 27, 55128 Mainz

Abstract. Human activities are exerting increasing impacts on the environment on all scales, in many ways outcompeting natural processes. This includes the manufacturing of hazardous chemical compounds which are not produced by nature, such as for instance the chlorofluorocarbon gases which are responsible for the "ozone hole". Because human activities have also grown to become significant geological forces, for instance through land use changes, deforestation and fossil fuel burning, it is justified to assign the term "anthropocene" to the current geological epoch. This epoch may be defined to have started about two centuries ago, coinciding with James Watt's design of the steam engine in 1784.

The Holocene

Holocene ("Recent Whole") is the name given to the post-glacial geological epoch of the past ten to twelve thousand years as agreed upon by the International Geological Congress in Bologna in 1885 (Encyclopaedia Britannica 1976). During the Holocene, accelerating in the industrial period, mankind's activities grew into a significant geological and morphological force, as recognised early by a number of scientists. Thus, in 1864, G. P. Marsh published a book with the title "Man and Nature", more recently reprinted as "The Earth as Modified by Human Action" (Marsh 1965). Stoppani in 1873 rated mankind's activities as a "new telluric force which in power and universality may be compared to the greater forces of earth" [quoted from Clark]. Stoppani already spoke of the anthropocene era. Mankind has now inhabited or visited all places on Earth; he has even set foot on the moon. The great Russian geologist and biologist Vernadsky (1998) in 1926 recognized the increasing power of mankind in the environment with the following excerpt "... the direction in which the processes of evolution must proceed, namely towards increasing consciousness and thought, and forms having greater and greater influence on their surroundings". He, the French Jesuit priest P. Teilhard de Chardin and E. Le Roy in 1924 coined the term "noösphere", the world of thought, to mark the growing role played by mankind's brainpower and technological talents in shaping its own future and environment.

The Anthropocene

Supported by great technological and medical advancements and access to plentiful natural resources, the expansion of mankind, both in numbers and per capita exploitation of Earth's resources has been astounding (Turner et al. 1990). To give some major examples:

During the past 3 centuries human population increased tenfold to 6000 million, growing by a factor of four during the past century alone (McNeill 2000). This growth in human population was accompanied e.g. by a growth in the cattle population to 1400 million (McNeill 2000) (about one cow per average size family). Urbanisation has even increased 13 times in the past century. Similarly large were the increases in several other factors, such as the world economy and energy use (see Table 1.2.1). Industrial output even grew forty times (McNeill 2000). More than half of all accessible fresh water is used by mankind. Fisheries remove more than 25 % of the primary production of the oceans in the upwelling regions and 35 % in the temperate continental shelf regions (Pauly and Christensen 1995).

In a few generations mankind is exhausting the fossil fuels that were generated over several hundred million years, resulting in large emissions of air pollutants. The release of SO_2, globally about 160 Tg/year to the atmosphere by coal and burning, is at least two times larger than the sum of all natural emissions, occurring mainly as marine dimethyl-sulphide from the oceans (Houghton et al. 1996). The oxidation of SO_2 to sulphuric acid has led to acidification of precipitation and lakes, causing forest damage and fish death in biologically sensitive regions, such as Scandinavia and the Northeast of North America. Due to substantial reduction in SO_2 emissions, the situation in these regions has improved in the meanwhile. However, the problem is getting worse in East Asia.

From Vitousek et al. (1997) we learn that 30-50 % of the world's land surface has been transformed by human action; the land under cropping has doubled during the last century at the expense of forests which declined by 20 % (McNeill 2000) over the same period. Coastal wetlands are also affected by humans, having resulted for instance in the loss of 50 % of the world's mangroves.

More nitrogen is now fixed synthetically and applied as fertilisers in agriculture than fixed naturally in all terrestrial ecosystems. Over-application of nitrogen fertilisers in agriculture and especially its concentration in domestic animal manure have led to eutrophication of surface waters and even groundwater in many locations around the world. They also lead to the microbiological production of N_2O, a greenhouse gas and a source of NO in the stratosphere where it is strongly involved in stratospheric ozone chemistry. The issue of more efficient use of N fertiliser in food and energy production has recently been summarised in a special publication of Ambio (2002).

The release of NO into the atmosphere from fossil fuel and biomass combustion likewise is larger than the natural inputs, giving rise to photochemical ozone ("smog") formation in extensive regions of the world. Human activity has increased the species extinction rate by thousand to ten thousand fold in the tropical rain

Table 1.2.1 A partial record of the growths and impacts of human activities during the 20ᵗʰ century

Item	Increase Factor, 1890s-1990s
World population	4
Total world urban population	13
World economy	14
Industrial output	40
Energy use	16
Coal production	7
Carbon dioxide emissions	17
Sulphur dioxide emissions	13
Lead emissions	≈ 8
Water use	9
Marine fish catch	35
Cattle population	4
Pig population	9
Irrigated area	5
Cropland	2
Forest area	20% decrease
Blue whale population (Southern Ocean)	99.75 % decrease
Fin whale population	97 % decrease
Bird and mammal species	1 % decrease

J. R. Mc Neill, Something New Under the Sun, Norton, 2000

forests (Wilson 1992). As a result of increasing fossil fuel burning, agricultural activities, deforestation, and intensive animal husbandry, especially cattle holding, several climatically important "greenhouse" gases have substantially increased in the atmosphere over the past two centuries: CO_2 by more than 30 % and CH_4 by even more than 100 % (see Table 1.2.2), contributions substantially to the observed global average temperature increase by about 0.5°C that has been observed during the past century. According to the reports by the Intergovernmental Panel of Climate Change in 1995 (Houghton et al. 2001): "The balance of evidence suggests a discernable human influence on global climate" and in 2001: "There is new and stronger evidence that most of the warming observed over the last 50 years is attributable to human activities". Depending on the scenarios of future energy use and model uncertainties, the increasing emissions and resulting growth in atmospheric concentrations of CO_2 are estimated to cause a rise in global average temperature by 1.4 − 5.8°C during the present century, accompanied by sea level rise of 9-88 cm (and 0.5-10 m until the end of the current millennium). Major anthropogenic climate changes are thus still ahead.

Table 1.2.2 Composition of Dry Air at Ground Level in Remote Continental Areas

	FORMULA	CONCENTRATIONS 1998 /pre-industrial	GROWTH (% YEAR) average (1990-1999)
Nitrogen	N_2	78.1 %	
Oxygen	O_2	20.9 %	
Argon	Ar	0.93 %	
Carbon dioxide	CO_2	365/280 ppmv	+ 0.4
Methane	CH_4	1.745/0.7 ppmv	+ 0.3 −0.5
Ozone	O_3	10-100/20 (?) nmol/mol	variable
Nitrous oxide	N_2O	314/270 nmol/mol	+ 0.25
CFC-1	$CFCl_3$	0.27/0 nmol/mol	< 0 (decline)
CFC-12	CF_2Cl_2	0.53/0 nmol/mol	< 0 (decline)
OH (HYDROXYL)	OH	$\approx 4 \times 10^{-14}$	?

Furthermore, mankind also releases many toxic substances in the environment and even some, the chlorofluorocarbon gases ($CFCl_3$ and CF_2Cl_2), which are not toxic at all, but which nevertheless have led to the Antarctic springtime "ozone hole" and which would have destroyed much more of the ozone layer if no international regulatory measures to end their production by 1996 had been taken. Nevertheless, due to the long residence times of the CFCs, it will take at least another 4-5 decades before the ozone layer has recovered.

Considering these and many other major and still growing impacts of human activities on earth and atmosphere, and at all, including global, scales, it thus is more than appropriate to emphasise the central role of mankind in geology and ecology by using the term "Anthropocene" for the current geological epoch. The impact of current human activities is projected to last over very long periods. According to Loutre and Berger (Loutre and Berger 2000), because of past and future anthropogenic emissions of CO_2, climate may depart significantly from natural behaviour even over the next 50,000 years.

To assign a more specific date to the onset of the "Anthropocene" is somewhat arbitrary, but we suggest the latter part of the 18th century, although we are aware that alternative proposals can be made. However, we choose this date because, during the past two centuries, the global effects of human activities have become clearly noticeable. This is the period when data retrieved from glacial ice cores show the beginning of a growth in the atmospheric concentrations of several "greenhouse gases", in particular CO_2, CH_4 and N_2O. Such a starting date coincides with James Watt's invention of the steam engine in 1782.

Without major catastrophes like an enormous volcanic eruption, an unexpected epidemic, a large-scale nuclear war, an asteroid impact, a new ice age, or continued plundering of Earth's resources by partially still primitive technology (the last four dangers can, however, be prevented in a real functioning noösphere) mankind will remain a major geological force for many millennia, maybe millions of years, to come. To develop a world-wide accepted strategy leading to sustainability of ecosystems against human induced stresses will be one of the greatest tasks of mankind, requiring intensive research efforts and wise application of the knowledge thus required in the noösphere, now better known as knowledge or information society.

Hopefully, in the future, the "anthropocene" will not only be characterised by continued human plundering of Earth's resources and dumping of excessive amounts of waste products in the environment, but also by vastly improved technology and management, wise use of Earth's resources, control of human and domestic animal population, and overall careful manipulation and restoration of the natural environment. There are enormous technological opportunities. Worldwide energy use is only 0.03 % of the solar radiation reaching the continents. Only 0.6 % of the incoming visible solar radiation is converted to chemical energy by photosynthesis on land and 0.13 % in the oceans. Of the former about 10 % go into agricultural net primary production. Thus, despite the fact that humans appropriate 10-55 % of terrestrial photosynthesis products (Rojstaczer, Sterling and Moore 2001), there are plenty of opportunities for energy savings, solar voltaic and maybe fusion energy production, materials' recycling, soil conservation, more efficient agricultural production, et cetera. The latter even makes it possible to return extended areas now used for agricultural to their natural state.

There is little doubt in my mind that, as one of the characteristic features of the "anthropocene", distant future generations of "homo sapiens" will do all they can to prevent a new ice-age from developing by adding powerful artificial greenhouse gases to the atmosphere. Similarly, any drop in CO_2 levels to excessively low concentrations, leading to reductions in photosynthesis and agricultural productivity would be combated by artificial releases of CO_2. With plate tectonics and volcanism declining, this is not a scenario devoid of any realism, but of course not urgent in any way. And likewise, far to the future, "homo sapiens" will deflect meteorites and asteroids before they could hit the Earth (Lewis 1996). Humankind is bound to remain a noticeable geological force, as long as it is not removed by diseases, wars, or continued serious destruction of Earth's life support system, which is so generously provided by nature cost-free.

Conclusions

To conclude: existing, but also difficult and daunting tasks lie ahead of the global research and engineering community to guide mankind towards global, sustainable, environmental management into the anthropocene (Schellnhuber 1999).

References

Clark WC, Munn RE (eds) (1986) Sustainable development of the biosphere. Cambridge University Press, Cambridge, ch 1

Encyclopaedia Britannica Inc. (ed) (1976) Encyclopaedia Britannica, Micropaedia IX. Encyclopaedia Britannica Inc., Chicago

Houghton JT, Jenkins GJ, Ehpraums JJ (eds) (1990) Climate change: the IPCC scientific assessment. Cambridge University Press, Cambridge

Houghton JT et al. (eds) (1996) Climate change 1995 – the science of climate change: contribution of working group I to the second assessment report of the Intergovernmental Panel on Climate Change. Cambridge University Press, Cambridge

Houghton JT et al. (eds) (2001) Climate change 2001, the scientific basis. Working group I contribution to the Third Assessment Report of the IPCC. Cambridge University Press, Cambridge

Lewis JS (1996) Rain of iron and ice. Addison Wesley, Readings

Loutre MF, Berger A (2000) Future climatic changes: are we entering an exceptionally long interglacial? Climatic Change 46:61-90

Marsh GP (1965) The earth as modified by human action. Belknap Press/Harvard University Press, Cambridge

McNeill JR (2000) Something new under the sun. WH Norton and Company, New York/London

Pauly D, Christensen V (1995) Primary production required to sustain global fisheries. Nature 374:255-257

Rojstaczer S, Sterling SM, Moore NJ (2001) Human appropriation of photosynthesis products. Science 294:2549-2552

Schellnhuber HJ (1999) 'Earth system' analysis and the second Copernican revolution. Nature 402:C19-C23

Turner BL et al. (1990) The earth as transformed by human action. Cambridge University Press, Cambridge

Vernadsky VI (1998) The biosphere. Translated and annotated version from the original of 1926. Copernicus/Springer, New York

Vitousek PM et al. (1997) Human domination of earth's ecosystems. Science 277: 494-499

Wilson EO (1992) The diversity of life. The Penguin Press, London

1.3   Sustainability Science

Jill Jäger

Sustainable Europe Research Institute (SERI), Garnisongasse 7/27, 1090 Vienna, Austria

Introduction

Sustainable development has occupied a place on the global agenda since at least the Brundtland Commission's 1987 report "Our Common Future." The prominence of that place has been rising, however. UN Secretary General Kofi Annan reflected a growing consensus when he wrote in his Millennium Report to the United Nations General Assembly that "Freedom from want, freedom from fear, and the freedom of future generations to sustain their lives on this planet" are the three grand challenges facing the international community at the dawn of the 21st century. Though visions of sustainability vary across regions and circumstances, a broad international agreement has emerged that its goals should be to foster a transition toward development paths that meet human needs while preserving the earth's life support systems and alleviating hunger and poverty. Science and technology are increasingly recognised to be central to both the origins of sustainability challenges, and to the prospects for successfully dealing with them.

Sustainability science

In recent years the contributions of the scientific community to the challenges of sustainable development have increased with reports of national and international groups, as well as from independent networks of scholars and scientists. To further discussion of the scientific challenges, two dozen scientists, drawn from the natural and social sciences and from all parts of the world convened under the auspices of an ad hoc organising committee at Sweden's Friibergh Manor in October 2000. Participants concluded that promoting the goal of sustainability will draw attention to new scientific questions, different research approaches and institutional innovations[1].

---

[1] The results of the workshop were published as: Robert Kates et al. 2001: "Sustainability Science", Science, 292, pp 641-642. http://sustsci.harvard.edu/keydocs/fulltext/2000-33.pdf.

Sustainability science is a response to the substantial but limited understanding of nature-society interactions gained in recent decades through work in the environmental sciences that factors in human impacts, and work in social and development studies that factors in environmental influences. But urgently needed now is a better general understanding of the complex dynamic interactions between society and nature. That will require major advances in our ability to analyse the behaviour of complex self-organising systems, irreversible impacts of interacting stresses, multiple scales of organization and various social actors with different agendas. Much contemporary experience points to the need to address these issues through integrated scientific efforts focused on the social and ecological characteristics of particular places or regions.

By structure and by content, sustainability science must differ fundamentally from most science as we know it. What were essentially sequential phases of scientific inquiry such as conceptualising the problem, collecting data, developing theories and applying the results must become parallel functions of social learning, additionally incorporating the elements of action, adaptive management and "policy-as-experiment". Familiar forms of developing and testing hypotheses are running into difficulties because of non-linearity, complexity and long time lags between actions and their consequences. All these problems are complicated by our inability to stand outside the nature-society system. Sustainability science will therefore need to employ new methodologies such as the semi-quantitative modelling of qualitative data and case studies, and inverse approaches that work backwards from undesirable consequences to identify pathways that can avoid those outcomes. Scientists and practitioners need to work together to produce trustworthy knowledge that combines scientific excellence with social relevance.

Meeting the challenge of sustainability science will also require new styles of institutional organisation to foster inter-disciplinary research and to support it over the long term; to build capacity for such research, especially in developing countries; and to integrate such research in coherent systems of research planning, assessment and decision support. We need to be able to involve both scientists and practitioners in setting priorities, creating new knowledge, and testing it in action. This will require systems to integrate work situated in particular places and grounded in particular cultural traditions with broader networks of research and monitoring.

Moving towards the World Summit on Sustainable Development

Taking the Friibergh meeting as a starting point, a number of regional meetings were organised in 2001/2002 to explore further the challenges of harnessing science and technology for sustainable development. A meeting was held in

Mexico City[2] in May 2002 to synthesise the results of these regional meetings and a range of related activities. This synthesis provided valuable input to the preparations of the World Summit on Sustainable Development, held in Johannesburg in August 2002.

The consultative process synthesised at Mexico City identified a rich variety of ways in which science and technology (S&T) has already contributed to sustainable development around the world. The consultations provided evidence of a wide range of science and technology based activities that, if vigorously pursued over the next five years, could yield tangible improvements in local and regional sustainability. Some of these activities involve the creation of new knowledge, others the better and more wide-spread application of knowledge that already exists. Which specific activities merit highest priority should be decided through consultation with affected stakeholders struggling with sustainable development action programs in particular places around the world. The Mexico City Workshop suggested a range of contributions that could reasonably be expected from the S&T community over the near term.

A "new contract" between science and society

While the relevance of S&T to sustainable development is generally acknowledged, a large gap persists between what the S&T community thinks it has to offer, and what society has demanded and supported. In recognition of this gap, the S&T community is increasingly calling for a "new contract" between science and society for sustainable development[3]. Under the contract, the S&T community would devote an increasing fraction of its overall efforts to research and development (R&D) agendas reflecting socially determined goals of sustainable development. In return, society would undertake to invest adequately to enable that contribution from science and technology, from which it would benefit through the improvement of social, economic and environmental conditions.

Making the "new contract" a reality will require changes in both the "demand" and the "supply" sides of science and technology for sustainable development. Increasing the demand for S&T will require increasing public and political aware-

---

[2] The results of the Mexico City meeting are published in: William C. Clark et al. 2002. "Science and Technology for Sustainable Development: Consensus Report and Background Document." ICSU Series on Science for Sustainable Development. ICSU, Paris. See also www.icsu.org; http://sustainabilityscience.org.

[3] See, for example, Jane Lubchenco 1998: "Entering the century of the environment: A new social contract for science." Science, 279, pp 491–497. Other references can be found in the background document for the Mexico City Workshop published in the ICSU Series on Science for Sustainable Development (see Footnote 2).

ness of the nature and magnitude of the challenges posed by transitions to sustainability. It will also mean convincing society that it can look to the S&T community for contributions to solutions and increasing the supply of contributions. This will require building the capacity needed to scale up those contributions adequately to address the magnitude of the sustainability challenges. Partnerships with all major stakeholders will be necessary, including the private sector, the public health sector and civil society. Indigenous and traditional knowledge must play a greater role in addressing sustainability challenges.

Implications for Research and Development

To become an attractive partner for society in the proposed "new contract," the S&T community needs to complement its traditional approaches with several new orientations. R&D priorities should be set and implemented so that S&T contribute to solutions of the most urgent sustainability problems as defined by society, not just by scientists. S&T for sustainable development needs to become an enterprise committed to empowering all members of society to make informed choices, rather than providing its services only to states or other powerful groups. Finally, given the inevitably unpredictable and contentious course of social transitions toward sustainability, S&T needs to see its role as one of contributing information, options and analysis that facilitate a process of social learning rather than providing definitive answers.

The substantive focus of much of the R&D needed to promote sustainable development will have to be on the complex, dynamic interactions between nature and society ("socio-ecological" systems), rather than on either the social or environmental sides of this interaction. Moreover, some of the most important interactions will occur in particular places, or particular enterprises and times. This means that S&T will have to broaden where it looks for knowledge, reaching beyond the essential bodies of specialised scholarship to include endogenously generated knowledge, innovations and practices. Devising approaches for evaluating which lessons can usefully be transferred from one setting to another is a major challenge facing the field.

For knowledge to be effective in advancing sustainable development goals, it must be accountable to more than peer review. In particular, it must be sufficiently reliable (or "credible") to justify people risking action upon it, relevant to decision makers' needs (i.e. "salient") and democratic in its choice of issues to address, expertise to consider and participants to engage (i.e. socially and politically "legitimate"). Evidence presented in the consultations suggests that these three properties are tightly interdependent, and that efforts to enhance one may often undermine the others. In particular, a simple focus on maximizing one of these attributes (e.g., is the science credible?) is an insufficient and counterproductive strategy for contributing to real world problem-solving where a mix of all three attributes is essential. The interdependence of saliency, credibility and legitimacy poses substantial challenges to the design of institutions for mobilising R&D, assessment and decision support for sustainable development.

The prospects for successfully navigating transitions toward sustainability will depend in large part on an improved dialog between the S&T community and problem solvers pursuing sustainability goals. Significantly, this needs to be done in ways that enhance the ability of problem-solvers at all levels to harness S&T from anywhere in the world in meeting their goals. It will be essential to understand what sorts of institutions can best perform these complex bridging roles (i.e. act as "boundary organisations") – between science and policy, and across scales and across the social and natural science disciplines - under a wide range of social circumstances. In addition, in a rapidly changing world of interdependence, such institutions need to be agile. There is a clear demand for systematic efforts to analyse comparatively the performance of experiments in the design of institutions for linking knowledge and action to identify how and under what conditions some "boundary organisations" work better than others, and above all to help the groups running the existing institutions to learn from one another.

Partnerships between the S&T communities and the private sector will be essential to promoting sustainable development, but forming effective partnerships is proving to be quite difficult. There is a particular need to effectively engage private sector scientists and engineers in multi-stakeholder efforts to address societies' most urgent problems.

The need for capacity building

S&T cannot effectively contribute to sustainable development without basic scientific and technological capacity. It is necessary to build capacity in interdisciplinary research, understanding complex systems, dealing with irreducible uncertainty, and to integrate across fields of knowledge, as well as harness and build capacity for technological innovation and diffusion in both the private and public sectors. The consultations showed particularly deep concerns about the shortage of science and engineering resources in developing countries and a decline of existing S&T in some countries. Science teaching at all levels must be enhanced, including efforts to "train the trainers". Efforts are required to support the mobility of scientists, to provide incentives for the development of a diverse technology community, to facilitate the participation of women. Exchanges of scientists and engineers are a proven method of capacity enhancement. Since in matters of sustainable development it seems that scientists and engineers in all regions of the world have something to teach one another, such exchanges must include South-to-North, as well as North-to-South and South-to-South dimensions. This will require building and maintaining the quality of key institutions of learning, provision of adequate infrastructure, and responding to the challenge of "brain drain". These requirements can only be met if appropriate strategies and policies are fully integrated in national development goals, including the enhancement of life-long learning, support for creative use of information technologies and maintaining S&T knowledge for sustainable development in the public domain. In addition, young scientists should be empowered to participate in developing the science and technology agenda, and there should be an increase in their number drawing in particular from traditionally under-represented groups.

Local versus global and the need for an empirical basis

A major conclusion from the consultations that were synthesized in the Mexico City workshop is that a great deal of the help that science and technology can provide to sustainable development must emerge from solution-focused R&D conducted in close collaboration with "local" stakeholders and decision makers. How "local" such collaborations need to be is itself a matter of some debate. But it is clear that agenda setting at the global, continental, and even national scale will miss a lot of the most important needs.

Notwithstanding the need for priority setting that reflects local needs, science and technology will be severely hampered in promoting sustainability until it has developed a much firmer empirical foundation for its efforts than is available today. A determined effort to move from case studies and pilot projects toward a body of comparative, critically evaluated knowledge is therefore urgently needed. In addition, progress toward sustainability will require a constant feedback from observations. Such observation provides a reference for theoretical debates and models on strategies for vulnerability reduction, and metrics for measuring success. In order to ensure a data stream needed to form the empirical basis of sustainability science, the observations of the natural sciences and of economic reporting should be augmented in the fields of socio-economic indicators, world views and society-biosphere interactions. An observation system for sustainability science will need to be based on a large sample of comparative regional studies, emphasising meaningful, relevant and practical indicators. Standards for documentation and access to data will also have to be developed. At least some of these foundation-building activities seem particularly well suited to the work of international science programs and collaborative efforts among the world's scientific academies.

Despite the need for a "place-based" or highly contextualised character of much of the science and technology needed to promote sustainability, the need to deepen and strengthen work on certain core concepts arose repeatedly in the consultations summarised at Mexico City. Many of these concepts were outlined at the Friibergh Workshop on Sustainability Science early in the consultations, and have been further developed in discussions on the Forum on Science and Technology for Sustainability (http://sustainabilityscience.org). Three topics, however, emerged from the Mexico City workshop as meriting special attention:

- Adaptiveness, vulnerability and resilience in complex socio-ecological systems
- Sustainability in complex production-consumption systems
- Institutions for sustainable development.

The funding dilemma

With a few important but relatively small and under-funded exceptions, efforts to "sustain the lives of future generations on this planet" still lack dedicated, problem-driven and solution-oriented R&D systems with attendant funding mechanisms for research and technology innovation. Further steps in supporting S&T for sustainable development will require restructuring of existing funding mechanisms at local, national, regional and global scales to increase funding efficiencies and synergies by supporting integrated projects that address multiple goals and involve diverse stakeholders, as well as substantial increases in investments in S&T.

Final remarks

The sustainable development challenge is urgent, and the potential contribution of S&T to meeting that challenge through participation in the design of more robust and adaptive strategies of development is clear. There appears to be a renewed commitment of the S&T community around the world to serve as an active partner in realising that potential. Living up to this commitment will require substantial changes in the way that scientists do their work: knowledge is more likely to be used if it is produced through collaborative processes that allow for greater participation in setting S&T agendas by social stakeholders; attention needs to be devoted to practical solutions as well as to conceptual understanding; progress will require integrated analysis of the complex interactions between nature and society; and some of the most important of those interactions for sustainable development take place at local to regional scales. Although S&T have made substantial contributions to sustainability goals, scaling up those contributions to a level commensurate with the magnitude of the sustainable development problem will require leadership in designing more effective communication between the S&T community and society; capacity building through education, the recruitment of the best young scientists and engineers world-wide to work on sustainability issues, closer collaboration with the private sector; and an array of innovative financing mechanisms.

Acknowledgments

This paper is based on the outputs of the Friibergh workshop held in October 2000 and the Mexico City Synthesis workshop held in May 2002. It thus relies on the input of a wide number of colleagues, to whom I am extremely grateful. Their numerous and excellent contributions to the discussions of harnessing S&T for sustainable development have been inspiring.

References

Clark WC et al. (2002) Science and technology for sustainable development: consensus report of the Mexico City synthesis workshop, 20-23 May 2002. http://sustsci.harvard.edu/ists/synthesis02/output/ists_mexico_consensus.pdf
ICSU (ed) (2002) ICSU series on science for sustainable development. ICSU, Paris
Kates R et al. (2001) Sustainability science. Science 292:641-642
Lubchenco J (1998) Entering the century of the environment: a new social contract for science. Science 279:491-497

1.4   What about Complexity of Earth Systems?

Horst J. Neugebauer

Geodynamics, University of Bonn, Nussallee 8, 53115 Bonn, Germany

Introduction

Life and environmental sciences are increasingly concerned about the appreciation of beneficial aspects of technology. There appears to be a concealed suspicion about the potential implications on life and environment. Science is thus addressing concepts on sustainability and control. Herewith research is moving to the modern concept of complexity and change. The nature of change has been addressed under the conceptual perspectives of objective entities and the dynamics of complex systems. The scale-concept appears to be a powerful tool for revealing the universal nature of change. The discussion is based on the failure of the concept of averages and the actuality of conditional relations.

Challenges of Management

The use of technical tools and patterns of behaviour for an improvement of living appears to be a common concept of life on earth. It is independent from the particular species or manifestation of life. We will not discuss this attempt in relation to a background of specific evolutionary progress or so, we just premise it as a common principal pattern of motivation of living beings. Especially human beings produced an exceptional development of tools and with the growth of natural sciences we became even more dependent on beneficial aspects of technology. Following the excitement of science and technology, we are thinking of even more fundamental improvements of the human life. This attitude is mainly rooted in and promoted by the pretension of the community of our interests. Consequently the request for potential adjustments of life and natural environment towards our appreciation through technology is raising more rapidly. Yet, at the same time there is a concealed suspicion about the consequences. Against this background we are getting inspired finding conditions of being protected from undergoing hurt, injury or loss; we are thus heading for controlling environment through sustainability concepts; in short, we try to find favourable prospects for a management of life and living conditions. Generally, the vision of management has both options, our individual mental base with the social society itself as well as our experienced environment and the respective resources. We might thus agree that, like for the progression of technology any potential success of management depends on the achievement of a precise assessment of our own interests. Presuming our technical tools are exciting undetermined changes to life and nature, we have no sound idea yet how to appraise environmental changes self-contained and properly.

According to our technical approaches we must admit that they are designed to operate like segregated well conditioned instruments, so called closed systems. In other words, they have been designed for an expected specific purpose or intention which make them originally depending on their causation. This narrow relationship is clearly built onto the reliability of the characteristic properties of such tools or objects. The general idea of managing nature independently becomes consequently promoted through the attempt of amplifying control. We are thus relying on that mere designing procedure, mostly ignoring the realm of change through mutual relations. In the case of our environment, we should principally not have overconfidence about its reliability from the beginning. Comparatively we find for the environment and likewise our own body frequently influential relations we have not recognised and experienced before. With respect to the supposed background of potentially unknown relations and the felt inability of properly addressing them, we would call this an open system. The actual boundaries of an open system such as our Earth appear to be likewise limited through our own interests, i.e. our present stage of respective knowledge and scientific abilities. This differentiation between the two concepts of anthropogenic and natural qualities seems to be quite common. It appears, moreover to be qualified for the purpose of investigating the qualities themselves from their distinction. Closed systems are thus characterised and limited as well by their assigned special purpose. They have to be preserved from failure through the maintenance of the originating supporting conditions and selected properties, such as a particular machine, a type of a car, a particular factory or an urban development. Therefore they seem to be controllable by reinforcing the expected properties through a continuation of the supporting conditions involved in their formation and function. While open systems perform ongoing changes, such as the most frequent transition between physical states of matter, but presumably on their own internal nature. For instance, the climatic system is operating on cross-relations incorporating the Earth and the solar system. Together they perform a differentiated spectrum of highly variable mutual interactions, but without any perceivable purpose by themselves.

We have now addressed some principal aspects concerning the intended polarity imposed through the separation between anthropogenic technology and our natural environment. Therefore we have to clarify the potential consequences of the imposed fundamental qualities with respect to their unacquainted mutual influential relationship. How much are closed systems subject to influences from the environment? Is there a fair chance to convey our limited knowledge from human technology to what we have called an open system? Seeking for an acceptable solution of the problem of discerning environmental change, the isolation of the anthropogenic impact onto the detached system has primarily been recommended. In a simplified but provoking manner we might thus embrace our cardinal problem in a straight way: How is change changing change? Before this puzzle is reducing us to despair we may first sally forth for a thorough check of our utilities which are supposed to have the abilities for a solution. Therefore we might go for some review of methods approaching complexity, the role of scales thereby and finally the occurrence and the quality of change in complex systems from a more physical and statistical point of view.

Normalcy versus Disasters

Science is commonly operating on approved techniques of observation and accustomed patterns of analysis and their preceding interpretation. This performance, however, is giving rise to another severe obstacle investigating Earth systems and the Anthropocene non distinctively. We might not hesitate to discover our traditional addictions of depending on approved standards and patterns both mentally and practically. One of those spiritualised "shortcuts" in science is the nearly ubiquitous usage of averages. Recording temperature, rainfall, water-gauges or deriving the amount of sedimentation; even investigating social behaviour or measuring material properties up to defining the state of matter; consistently we are reverting to the same pattern: Change becomes persistently ignored in our established perception and misrepresented quantitatively by averages. On the other side, recordings of rather eminent changes exceeding the ordinary are indicated as exceptional. However, changes are evidently independent from the perceived scale of experience as we will see later on; they are a common feature at any achievable scale of resolution. Well, we may possibly become amused by such a general statement and our excessive interests, the feasibility of change. Yet, as already mentioned, averages are unfortunately strongly dependent on the range of sampling and the quality of the appraisal. Clearly, our ignorance reflects primarily the addiction to make phenomena feasible more easily, and secondly it might express a common mental desire for permanency. So we will not be surprised to recognise the familiar concept of persistency which we equate with normalcy in contrast to reputed exceptional instances or events, the disasters. Not necessary to emphasise that such a simplification will evoke another inveterate shortcoming of our approach, frequently expressed by the sentence: the exception is proving the rule. Finally, we end up with the common concept of linear causality, whenever a characteristic property, i.e. a fixed average, becomes related to a causal relationship representing an expected behaviour. Such an idealised linear approach is comprehensively expressed by the well known statement: equal causes will excite equal consequences. Only the validity of this presumption of linearity becomes accountable to handle mutual influence within a system as a linear accumulation between independent constituents. So we have severe doubts, because of our experience, to separate and isolate phenomena from complex relationships such as an Earth system; neither our human influence from the environment nor likewise disasters from normalcy. What we have skipped implicitly through withdrawing ourselves onto mere averages is the signature of change. This means more specifically, the indication of the irreversible influence of further variables upon our records. Following the latter, we rather will find, cross to the upper principle, that: similar causes will not necessarily always excite similar consequences. Finally, we might agree, that we are not moving on solid ground for the separation of phenomena. There is urgent need for investigating the nature of change more deeply, gaining a qualified concept for an reliable distinction between reversible and non-reversible approaches of phenomena. Scales provide such an independent physical-statistical tool; they may shed some light onto both the revealing the idea of complexity and an evaluation of the concept of averages, respectively.

Scales and Complexity

Everybody is well acquainted with the notation scale, whenever someone is talking about maps. Scales are used here as a common and independent level of representing spatial features of selected phenomena. Generally, phenomena or properties of a certain size appear in a spatial context to each other on a distinctively scaled map. Independent from our choice, objects will either be composed by subordinate details on smaller scales. They may even become constituents of larger phenomena, when we are moving to sub- or super scales compared to the declared reference scale. Soon we find our selection of objects being conceptual and depending on our intrinsic interests. So we choose specific structures, for example an area of rare sedimentary deposits being reflected alternatively by some property or characteristic measure such as its diameter, a layers thickness or its square dimension. We would take this quantitative spatial property for a representative spatial scale, representing the qualitative phenomenon of a sedimentary basin for instance. As already mentioned, complementary information from the smaller or larger scale context is likely to be omitted, as in the case of the specific map. Supplementary we assigned to our addressed object a physical context, which led synonymously to a functional relationship. Here the information assigned to smaller scales becomes summarised in terms of fixed averages of properties. While the context on larger scales becomes usually more or less condensed into a regional source function, such as the course of subsidence and sedimentary loading regarding our example. With respect to our chosen reference scale, we thus integrate an inner and outer scale-context of attendant constituents into a linear causal relationship, as mentioned before. Thus, averaged material properties, for instance, become assigned to particular composite suites of layers, while the formation of the averaged layer is linked to an input function, which comprehends the influence of a surrounding relief with a highly variable soil- and drainage system. Thereafter this concept is comprehending an entire hierarchy of scales through a linear, reversible relationship, which is built onto one single, pretentiously representative scale. So far scales expose the general weakness of our conceptual approach based on averages and the idea of an objective isolation through assigning properties approximated from the context. Well, presumably we were primarily attracted from the steady attitude of the linear relations which actually signifies that change is only and uniquely due to the variation of a particular source function or causation in time. Yet, what should happen under this perspective with the entire zoo of related scales from the context we indicated before? In order to keep the linear functional relation alive, i.e. independent, we must claim that the entire spectrum of variation of properties of the scale-context erase each other at any subsequent moment. We may feel now quite disappointed about any accomplished linear translation of actual as well as historical records, depending either on time or spatial structures.

Let us therefore give the complex scale-context a trail. Watching change at a contemporary sedimentary basin for instance, one can never claim an independence of the respective filling procedure from the actual state of the basin itself, of the rainfall from the actual state of the lake, the river load from the actual river flow and bed shape and so on. Obviously, change appears rather as a result of a contemporaneous causality among all constituents. Thus a sedimentary record represents the irreversible sequence of momentary events of change, depending on contemporary complex interactions. This will cause severe implications for the presentation through the scale-concept. First of all, the entire scale-hierarchy is reflecting the phenomenology of a complex system. Second, the scale-context with respect to an arbitrary reference scale is expressing mutual interactions among the constituents. Third, there is no a priory preference of particular scales within the bounds of a complex system. Fourth, scales become representative of temporary events or instances in a complex system rather than for lasting objects. Fifth, a recorded scale, respectively event of change, is exclusively signifying a momentary state of the entire conditional relations. Finally, there seems to be ample evidence for a mutual dependence between variables of a complex system. The use of scales, respectively spatial properties for characterising phenomena enables a close follow up of the controversy about conceptual representations on a unique basis. The two opposing views are on one side the very familiar concept of steady objects related to a particular scale, motivated by an assigned expectation on its function. On the other side we become strongly attracted by the concept of complexity, which comprehends arising and ceasing of phenomena on the basis of multiple-scale conditional relations. Although we feel more attracted by the complex view, the response of a complex system can no longer be related to any particular causation. To benefit from the scale concept, we will turn back to some analysis and end with statistics of observations.

Anthropogenic versus Natural?

Three fundamental interests of research on life and environment are closely related to our preferential custom for averages or equivalently the addiction to the viewpoint of an objective, i.e. scale-related reality: these are the aspired identification of the human influence on an environment. Secondly, the increasing interest in an isolation through management of beneficial phenomena and finally, our striving for an access to the alleged exceptional nature of disasters. We may find the motivation of the interests themselves being already the cardinal source of the obstacles for a smooth solution. The previous chapter gave us an idea about both the weakness of the determined objective perspective, but even so a glance at the challenges of a conceptual complexity with the expectation of random behaviour. So, why can we not just proceed in the common manner? Answering this question gives some kind of an universal advice for our following discussion. We already have revealed statistical averages being commonly adopted for characterising objects implicitly as well as explicitly. This fundamental approach is establishing the conceptual view of so called representative scales or properties. However, whenever a representative scale is supposed to signify an object, the entire remaining scale-context has somehow to be compensated or rather eliminated. This is nothing else but the formation of an object through an intentional isolation of a particular property from its environment, which has to be achieved somehow. Technology is exercising this process of isolation through forceful conditioning of parts of the context; as we know, with limited success. A more principal approach becomes obvious from mathematics. Performing the calculation of averages, we are expecting a few dominant properties in a centre surrounded by an assigned assemblage of smaller and larger deviations. Such a statistical distribution is commonly approximated through the famous symmetrical Gaussian-function with its limited variance. The imposed concept is automatically entailing a total mutual long term balance of any deviations up to infinity, according to the ideal signal assemblage, the so called white noise. Two implications assigned with this concept are of striking importance for our considerations: assuming a white spectrum for the fluctuations around the mean, their originating system variables have to be independent from each other by definition. This could have been the aspired approval for all of our addressed themes; the isolation of objects, their linear additive attitude and the exceptional nature of extremes, the disaster. However, white noise appears to be an ideal, inaccessible theoretical concept only, which can neither analytically nor practically become materialised. And there are good reasons in addition, why we fail to reach the conditions of white noise eventually. Any approach is missing the extremely large constituents ranging up till infinity. Here we may remember that our experience of space is critically depending upon the perception of objects. Due to this fact, space remains always bound, it can merely be "unfolded" through our sensed experience. Well, because our final "hardness test" of an objective entity of phenomena is rather "leaking", we are drawn back again to the realm of complexity.

In particular we may derive two important implications from the previous analysis: first, variables performing fluctuating system behaviour ought to be depending on each other. Second, there is no real obligation for random fluctuations of a system to counterbalance each other. So far these consequences translate into our fundamental research interests in the following way: Changes performed within a coupled system can never be traced reversibly or become merely detached from a system because of the non obligatory compensation. Sustainability of technically designed phenomena will be even less than the applied conditional effort because of the persisting conditional relations. Without the pretended narrow range of counterbalance, extreme fluctuations of a complex system can principally not at all be ruled out. Well, the whole bunch of highly putative inventory of the concept of objective entities might not be rather encouraging for our endeavour. Yet, alternatively the approach by complexity needs to be investigated on a broader level, especially the aspects of unexpected change compared to what we have considered so far.

Complexity and Change

Complexity signifies primarily the investigation of change of phenomena, their variability and sequential appearance, comprising the conditional relations for their arising and ceasing. It seems not to be the right place here for discussing the aspect of theoretical modelling in detail. Yet, the inherent character of the complexity may become explicitly revealed through a statistical representation. Most instructive for our purpose will thus be the statistical analysis of the probability distribution on the occurrence of specific properties, such as length, size or intensity. Properties are indicating events or states, which we identified previously by scales. In particular those scales will still signify our weak conceptual objects but here we like to learn from the statistical approach to fluctuations something about the latent contextual relations. Those will materialise through an expected scaling rule of fluctuations addressed either in spatial qualities or sequential in time. In order to get the procedure affirmed, it would be really essential to sample properties in both respects an extremely large number and over a respectably wide range of sizes as well. Many quite valuable statistical compilations have been performed successfully such as for the probability distribution of earthquakes, land slides, drainage networks as far as to the tiny scales of pore space. At that point we clearly search for both the significance of irregularities of apparent scales as well as for signs, identifying concomitant constituents of an underlying system allowing instantaneous mutability. Two persistently conformed end members of probability distribution functions have been identified. They are represented in a logarithmic frame in Figure 1. Too narrow specified attempts of sampling, either in respect to the objectives or the time span of sampling leads purposefully to an old acquaintance: the normal distribution, representing a mean value with a limited statistical variance. Statistical probability approves now, that our pretended objects are just conceptual entities which can be experienced, but will principally not be separable from an observer. We will try quickly to illustrate that by means of a simple perceptual analogue. Let us survey the manifestation of life on Earth. Nobody would probably mind to characterise his body for a moment by just measuring its size. A comprehensive size statistics of the human species would thus result in a scale-dependent or normal distribution function, as shown in Figure 1.4.1. Personally one may appear above or below the average, but most likely within the limitations of the variance. Extremely small or large sizes are indeed not significant at all. Well, reflecting the fish and the steak we just had for dinner, our indigestion and probably the infectious disease we just passed some time ago, we might become aware of our tremendous dependence on even other manifestations of life. Feeling ashamed, we quickly give way to an extension of the statistics. In between the size of bacteria and that of whale, we most presumably will end up with some kind of distribution function of the power law type, as shown above. At the least, we will now recognise from the character of the straight line with the slope, there appears no longer any scale-dependent mean, not for any of the sampled species in that zoo of life forms. We finally lost our pretentious objective significance within the mutually depending life forms on Earth. Our little thought experiment illustrates complexity; the power law distribution represents the peculiar statistical quality of scale-invariance.

Considering, for instance, the strength of earthquakes with respect to the size of fractures, one finds the power law reconfirmed. Although our size statistics may

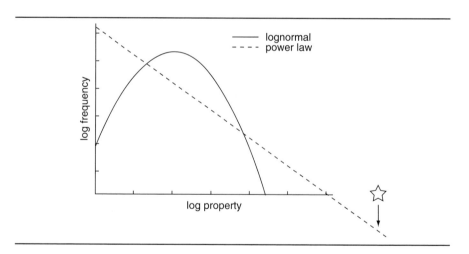

Fig. 1.4.1 Two statistical probability distribution functions in a logarithmic presentation. The normal distribution indicates much lower probabilities for small-scales and even less large-scale properties compared to the straight power law function. The normal distribution shows also excessive frequencies around the statistical mean value. Exceptionally large-scale events or instances, denoted here as disasters, are indicated by the star symbol.

cover a basket of semantic varieties ranging from microscopic-scale atomic friction events via fissures, micro-earthquakes, perceivable regular seismic events up to the destructive main shocks and disasters. The power law as their scaling rule will not reflect any of those specified semantic distinctions at all. Just the opposite, scale-invariance affirms an unrestricted equivalence between all of the sampled scales. There is not even a preference to be found for any particular pattern of succession for their occurrence in time. It appears thus a very reasonable conclusion which can be derived from the power law: there exists no other reality in complexity but change. Following such an universal rule of equivalence, we may return first to the spatial aspects of the probability distribution. Because of the slope of the power law distribution function, the equivalence operates in between impressively higher probabilities for small scales compared to the much lower ones for large scales. A measure for the quality of the equivalence of a particular phenomenology such as fractures, slides or layering becomes expressed through the slope of the function. This measure is indicating a spatial peculiarity, the fractal dimension. The latter classifies our properties being fractals, i.e., objects that are looking the same on all scales; they are denoted as statistically similar. Second, through the scale-invariance change is a priori independent from the scales selected through our perception or any other conceptual expectations considering complexity. Vice versa, it seems not meaningful to neither distinguish complexity statically as local, regional or global, nor assigning complexity with apparent extreme properties somewhere. Change becomes generally noticed as a subject of comprehension, whenever we specify any of our interests or experiences conceptually. Through this we will contribute to the realm of complexity, i.e., we increase a phenomenology and therewith the manifestation of space according to the statistical distribution function. This is leading us directly to a third, rather universal quality of space perception. Respecting scale-invariance, there appears a pronounced equivalent probability of smaller scales compared to the insignificant one for larger scales. Yet, the total range of properties and assigned space is generally bound but in particular indefinite; so bounds are just within the compass of our own mind. On one side, there are our limiting abilities to contemplate omnipresent minimum-size changes, on the other side we are operating with an exceedingly poor experience of extremely improbable properties or events. Obviously, both aspects together are consecutively leading to alterations of the bounds of complex systems and associated space perception. At the same time they are misleading us to defining objects through conceptual averages. Finally we are able to appreciate that in front of the fundamental statistical disposition of complexity: mere change, independent from any particular property.

So far, the scale-invariant and the scale-dependent distribution functions have been considered as two exponents of conflicting conceptual perspectives. The power law probability distribution appears to be well established through the entire range of conditional relations, constituting complexity. With this respect the representation of complexity is the actuality, the ultimate reality of omnipresent change. Thereafter any representative scale specification according to the normal distribution function has to originate from the complex relations. Conceptual objects are thus compulsory dependent, representing in every particular the potentiality of occurrence as event or instance. The latter "elements" of complexity thus obey a twofold nature: a mind-based or semantic notation, expressing an assigned quality which is originally linked with some manifestation from experience. Because of the inevitable universal change of all particular phenomena, they are necessarily maintaining complexity in return. Actually, our two conceptual perspectives of objective entities and fluctuating events form a closed generative loop of change; or they may simply be seen as two aspects of the ongoing dynamics of change. According to this interwoven background of change, we may create a multiplicity of phenomena without gaining the capability of neither foreseeing their change nor revealing the conditional relations of change. Finally we may try to translate this relationship back into phenomenology. According the potential for particular dangerous freak-waves at the ocean is tracing back to the same actual complexity as the frequent smooth undulations at the surface out at sea. In the same non-preferential manner we might regard examples such as an unusual storm beside the common gusts of the wind, an abnormal flood beside the familiar runnels or a shaking earthquake beside the crackling of micro-fissures.

Dynamics of Change

Although we are increasing complexity ourselves through differentiation along with intentional research and technology, our abilities of understanding complex systems remain still rather enigmatic. Yet, we may remember from the previous discussion that: complexity is built up by perpetual processes of ongoing change; the constituents generating change are not really independent from each other; those systems will thus generally be non-reversible. Last, but not least, supporting complex systems conceptually, we have no real choice of avoiding their consequences. Model-analogues appear to be very useful for learning about the implications of some fundamental rules according to the previous statements. There are quite reasonable approaches to complex system behaviour by means of locally-based, small-scale quantitative, numerical models. For example, spring-block models and cellular automata will at least reproduce the previously addressed probability distribution for fluctuating change. For another small thought-experiment, let us assume a model environment of a large number of local numerical supporting points. They are all communicating among their respective nearest neighbourhood, either physically through connecting springs or just via prescribed universal rules on the handling of incoming and outgoing information. In order to simulate the conceptual linear causality, we may force a particular limited section of the operating system to perform, for instance, regular sinusoidal changes. Nobody will hopefully expect any longer, that this spatially bound forced action remains unnoticed by the linked neighbours outside of the activated area. Let us proceed a step further, now the same spatial sub-district should become affected along its boundaries by two different, independent variables. While those are linked to each other only internally through some respective variable parameters. According to the applied conditions and the additional internal degrees of freedom, the sub-system will either respond directly to the external steady forces as before, or it may alternately reach internally some indifferent stages of vigorous pulsation. Compared to the previous situation, the sub-system appears no longer to be entirely controlled from the outside. Through that example we may get a small glance how external control through prescribed causation is successively lost to a complex system through an internal interdependence of variables. Therefore one finds the rules of neighbour to neighbour communication being of paramount importance in complex systems. As a consequence of that, we meet at any supporting point and any successive step through time a complex variation of influencing forces. Finally, we will perform a perturbation of the entire complex system by at least to external sources. According to the high number of local degrees of freedom and the independent internal variation of forces we may expect a fluctuating response of the simulation. That means, with any step of redistribution of the internal state of conditions, the system changes abruptly between responding properties. These simulations give us concomitantly an idea on the prominence of conditional relations and the distinguished quality of complexity. Change becomes compulsory just at the moment, when all of the related conditions are reached. Otherwise the dark and heavy clouds on the sky will pass by without giving birth to any expected drops of rain.

One might raise the question now on what we usually call the efficiency of complex systems measured by a comparison of the delivered change with the supplied energy. With the indicated weakness of averages and the assigned concept of objective entities we are limiting the applicability of the common physical principles of conservation in the long term perspective. This explains on one side the practical limitations of prediction in the linear approach. At the same time, however, it points towards an availability of excessive energy in the concomitant context of complexity. As this latent energy will never be accessible in its entirety, we may miss the opportunity for prediction of complex systems on conservation principles as well. Simulations of fluctuating changes according to the power law suggest a stable distinguished critical state of the complex systems. That means a continuous optimum state of both external energy support assigned to the constituents of causation and the internal energy represented through the constituents of complex reactions. Whenever the internal energy of the complex reactions is insufficiently supported, the scaling law of change is approaching a "diffusive" normal distribution. Amazing enough, in the complex frame of models we meet scale-independent change by the arising and ceasing of events according to the conditional play of external and internal energies. While conceptual scale-dependent objects arise through or perceptual attention and cease through our ignorance of the complex realm of change. Change is in any of our perspectives, so it depends on our personal addiction whether and where we take up this experience.

Is there a Message?

Complexity seems at the first moment not being a real technical term of improvement in order to make progress faster, cheaper and sustainable. Complexity may appear from that point of view to be rather another mere inconvenience, an unintended dubious obstacle. However, the benefit from considering complexity in our respect concerns in the first order an improved appraisal on what we are currently doing and possibly should avoid to do in future. To start with the failure of conceptual averages, we are losing the pretended independence of variables and thus the independent entity of objects. This implies complex relations and therewith change to be consecutive to any action without further particular proof. Secondly, complexity materialises in corresponding particulars and denotations, but with respect to the assigned change it remains an immanent and universal principle. Change appears thus to be the universal rule of reality with a plenitude of painful implications. Because of the related conditions we find thirdly, complex change is occurring abruptly with most events or instances. Precursors are not compelling due to the independence of change of any particular scale or property. Fourth, the mutual linkage between the conceptual perspectives of particular phenomena and their complex phenomenology provides the entire actuality of unpredictable change. Thus any striving, no matter how tiny it may be, will indeed be an integral part in the realm of complex change. Because of the equivalence of change, found from the fractal distribution, there is neither an excuse for an omission of small-scale, local activities, nor a good reason for a proclamation of preferential global moves only. There is just merely an equivalence of change. However, this elementary quality of the equivalence is revealing a perpetual increase of the actuality of change through any progressive split up of our potential objects. Any additional differentiation causes more "boundaries", increasing properties, scales and thus the growth of the potential of change. Well, boundaries appear to be the cardinal problem of change, the problem of safety and the problem of sustainability. We are the designers, we might thus reduce boundaries, especially those between man and nature.

Part 2

Global Change and Human Security

2.1 Climate Change, New Weather Extremes and Climate Policy

How to minimise the risk for global development?

Hartmut Graßl

Meteorological Institute, University of Hamburg and Max Planck Institute for Meteorology, Bundesstr. 53, 20146 Hamburg

Introduction

Climate is an important natural resource that varies on all time scales from months to billions of years. As the synthesis of weather and its statistics over a certain period include the weather extremes in the wings of frequency distributions. These extremes govern our infrastructure. Codes for construction of houses, the height of dams and dykes as well as other security infrastructure have been adapted to historical observed climate variability. If the frequency distributions of climate parameters like temperature and precipitation vary due to climate change our infrastructure is no longer well adapted and more frequent weather-related catastrophes will occur.

In the following I will firstly report about the observed changes of climate in the recent past, e.g. temperature increase, glacier retreat, sea level rise and also changes in variability of precipitation. Then the arguments for new weather extremes due to climate change will be introduced before I speak about changed extremes of the recent past and the lack of scientific rigor for the introduction of the so-called 100-year event. In scenario calculations future potential changes of extreme precipitation will then be projected that will lead me to a discussion of political measures taken so far by the United Nations, the proper body for reaction to a global problem.

This will lead us necessarily to a debate on a future energy system for mankind. As the climate system responds in delayed mode to disturbances I will finally discuss how the proper mix could look like between local adaptation to no longer avoidable climate change and global action to dampen the anthropogenic climate change rate; the latter rapidly enough to avoid a dangerous interference of mankind with the global climate system that would certainly threaten our approach to sustainable development.

Observed Facets of Climate Change

There is no longer any doubt that the 20th century saw a rapid mean global warming at the Earth's surface of about 0.6°C that occurred in two major steps, the first until about 1940 and the second since the late 1970s (IPCC 2001a, b). Since the total amplitude between a glacial epoch (the last one ended about 18,000 years ago) and an interglacial (often named warm period) is 4 to 5°C for the global mean near surface air temperature and needs thousands of years, such a warming within one century is extraordinary and has not occurred during the last 1000 years at least on the Northern Hemisphere, for which a data base from paleo indicators exists. Strongly above average warming has been observed over large parts of Siberia, Western Canada and Alaska, in agreement with the positive snow albedo – temperature feedback that is also a major agent when recovery from a glacial is observed. When a snow surface with a typical reflection of 60 to 80 percent is replaced by bare soil or vegetation the reflection shrinks to between 10 and 20 percent. Up to four times more solar radiation is then absorbed that additionally heats the surface.

The warming is less than average or even absent over ocean areas with strong mixing, because any warming will be delayed there due to the high heat capacity of the well mixed upper few hundred meters (please remember that the heat capacity of a 3 m ocean layer is equivalent to that of the entire atmosphere above).

For most people, especially those in subtropical and tropical climates, precipitation changes are more important than temperature changes. Therefore, the knowledge of precipitation trends would have an even stronger impact on our planning, e.g. for industry, agriculture and tourism. However, the routine measurement of precipitation is accurate enough only for stations with rainfall at low wind speed. Areal precipitation, e.g. for a distinct river catchment, cannot be derived reliably even from the best of these stations, as the network is nowhere dense enough for that purpose. And remote sensing via radars, fully covering an area, is far from being long and accurate enough for any trend analysis. Despite all these drawbacks we do have trend estimates for some stations, allowing a rough image of the major change under way: More precipitation in higher mid-latitudes, especially in the winter half year, a drying in many semi-arid areas of the subtropics and tropics, more in high northern latitudes throughout the year, and often a positive but significant change in the inner tropics (IPCC 2001a).

An obvious result of the observed mean global warming of near surface air is the frequent retreat of glaciers. Using data from all glaciers with a continuous mass balance observation in many but not all major mountain ranges, Haeberli et al. (2001) could derive a mass loss that is equivalent to about 30 cm per year in thickness. An example for the variability of glacier retreat: The European Alps have lost about 60 percent of their ice mass since 1850 but some glaciers in Western Norway advance, despite the warming also observed there, because winter-time precipitation has simultaneously increased by up to 30 percent in the 20th century, especially in recent decades.

Global mean sea level rises. Estimates derived largely from the same gauges at coastal stations by different groups range from 10 to 20 cm for the 20th century. Using in addition satellite altimeters since 1992 a rise of 2 mm per year is given as the best estimate for the average sea level rise during the recent decade. One of the main contributions to this rise is thermal expansion of seawater but also the glacier retreat contributes substantially (IPCC 2001a). Whether the large ice sheets over Greenland and Antarctica have contributed is not yet clear. The measurement of the geometry of these large ice masses, their volume being equivalent to about 70 m sea level rise, is still inadequate for that; however, it is already clear that the present net contribution to see level change must be rather small.

Why must climate change lead to new weather extremes?

Weather extremes are the rare events existing in the tails of frequency or probability distribution functions of climate parameters. As clearly seen in Figure 2.1.1 a shift of the mean value to higher values at time t2 compared to time t1 will make a certain large positive deviation from the mean much more frequent (middle hatched area). An example for summer-time temperature in England may help to underline this statement. Would temperature increase by 1.6°C in the summer half

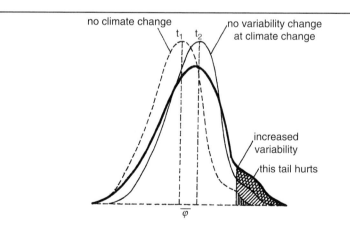

Fig. 2.1.1 Schematic representation of changes in the frequency distribution of a meteoroloical parameter caused by climate change. Even if the distribution (variability) does not change at time t2, new weather extremes must be observed on one side (hatched portion). If variability increases, as observed for precipitation in many regions, rare earlier events become much more frequent and new absolute extremes will be observed (double-hatched area); taken from Grassl (2001)

year until 2050, which is a frequent result of climate models for business as usual scenarios of human activities, and would the frequency distribution not change, then today's 3-σ event with a 1.3 percent chance of occurrence would become a 1-σ and thus have a frequency of occurrence of about one third, i.e., be 25 times more frequent. If the frequency distribution becomes broader with the shift of the mean, what is observed for the amount of precipitation per event, the increase in frequency of occurrence would be even larger (see double hatched upper area in Fig. 2.1.1). In summary, any climate change must make weather extremes more frequent at one side of the distribution and in most cases less frequent at the other. However, as our infrastructure is adapted to the "old" distribution, we will experience more damage, and also new extremes must occur. How we should react in this difficult situation will be the topic of section 8.

Observed Changes in Weather Extremes

Weather extremes are rare events. Therefore, the detection of changes of these rare events is a very difficult task. The first prerequisite is long time series that exist only in some regions with sufficiently high accuracy and the second is unchanged observing instruments and methods or – if changes occurred – a joint period of measurement for both the old and new instrument or evaluation method. These conditions restrict the search for changed frequency distributions to only few parameters, namely station air pressure, air temperature, daily rain amount and river level over continental areas mostly in Europe, North America and South and East Asia.

For example, a 120-year time series of sea level pressure observations at three coastal stations in the German Bight allowed the derivation of storm frequency with relevance for surges in the Elbe estuary. The result: Storm frequency (wind strength above Beaufort 8 from a direction causing storm surges in Hamburg) shows two broad maxima in 1920 to 1930 and about 60 years later but no significant trend (v. Storch and Schmidt 1998). A search for changed frequency of floods in few tributaries to the rivers Danube and Rhine in Southwest Germany that are not strongly affected by local human activities revealed a massive increase in flood frequency. What earlier was named "flood of the century" with a 100 year average return period in the first half of the 20th century became a 20 or 10 year flood in the second half of the century (Caspary and Haeberli 1999). The reason: increased total wintertime precipitation but not strongly increased precipitation per event during the recent decades (see also Box "The 100 Year Event": a shaky fundament).

However, the most important finding in the search for changed frequency distributions of climate parameters is the following:

In all areas with an increase or a stagnation of total annual precipitation during the 20th century and in many areas with slightly decreasing total precipitation the amount per rain event increased, i.e. severe precipitation became more frequent (IPCC 2001a). This observation is consistent with the Clausius-Clapeyron equation that states: For a temperature increase of only 1°C the maximum possible water vapour amount in air (saturation pressure) increases by 6, 8, 20% at 20, 0 and −80°C. Therefore, higher surface temperatures must lead to higher maximum precipitation rates at otherwise unchanged circulation conditions (see also Box "Gedankenexperiment").

In summary, it is the higher probability for severe precipitation events in a warmer world, which will cause most of the threat to our infrastructure. We will need larger sewers, re-enforced and partly relocated dams and dykes.

Projected Changes in Climate Parameters and Related Weather Extremes

As pointed out by IPCC (2001a), the observed near surface temperature increase in the 20th century, especially during the last 3 to 5 decades, is with high probability due to human action. Therefore, climate model projections of further warming in the 21st century have recently got much more attention. Many scenarios of climate change exist (so-called SRES scenarios of IPCC (2001a)) leading to a wide span of global mean warming at the surface, largely depending on assumptions about population increase or its stabilisation, global mean economic growth, degree of globalisation, speed of technological advances and preferences for certain energy sources. None of these scenarios was explicitly driven by climate policy although some come close to such scenarios, which are also available in literature but often use prescribed maximum CO_2 concentrations in the atmosphere without assessing the consequences for economics.

There are only a few attempts to derive extreme weather events from climate model runs. The main reason for the reluctance to try it is the lack of spatial resolution in long model runs, as weather extremes are often local or regional in scale and these small scale phenomena often cannot be described in global coupled atmosphere/ocean models.

With increasing computer power more attempts will be started and we may soon see the results. One of these attempts by Semenov and Bengtsson (2002) will be mentioned here.

The authors show that the observed increase in rainfall per event will continue if warming at the surface continues. This intensification of the water cycle is probably the main physical threat related to climate change for inland areas especially in mid and high latitudes. For coastal zones sea level rise is an additional and – maybe – very often the largest impact. What are the consequences of an intensified water cycle? First of all a higher frequency of flash floods but also one for major wintertime flooding in mid-latitudes. In a further attempt to derive the physical consequences of global warming by an enhanced greenhouse effect for Europe, Palmer and Räisänen (2001) asked the following question: Will there be more wet winters? As Figure 2.1.2, taken from their publication, depicts Central and Northern Europe will get much more wet winters (defined as surpassing one standard deviation). For citizens, authorities and insurance companies this is rather bad news, because winter floods, e.g. along the rivers Rhine, Elbe, Odra, Vistula, will become

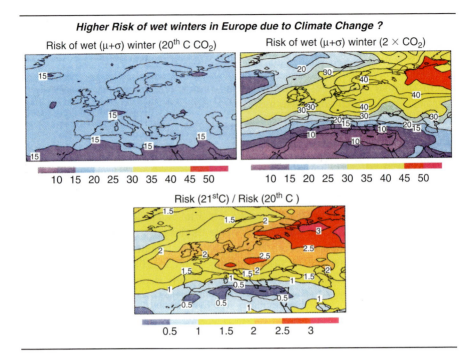

Fig. 2.1.2 Change in the probability of a wet winter defined as the mean plus one standard deviation. Upper left: present situation; upper right: probability in percent during a transient coupled model run at the time of CO_2-doubling (years 61 to 80 from present); bottom: ratio of both panels giving the change in risk of a wet winter arising from anthropogenic climate change (business as usual scenario), from Palmer and Räisänen (2001)

more frequent and snow cover duration will be drastically reduced. The author's recommendation for services and authorities is: please evaluate existing time series to find tendencies of variability of climate parameters and derive from these findings extrapolations (probable futures) for the coming decades, and adopt them as basis for changed infrastructure (often enforced protection).

UNFCCC and the Kyoto Protocol

When the first full assessment report of the Intergovernmental Panel on Climate Change (IPCC 1990) had appeared and had been discussed at the Second World Climate Conference (SWCC) in Geneva, Switzerland, the ministerial part of this conference requested in early November 1990 a framework convention on climate change to be ready for signature at the United Nations Conference on Environment and Development in Rio de Janeiro, Brazil, in June 1992. The United Nations Framework Convention on Climate Change (UNFCCC), which has been signed in

Rio by 154 countries, was resting on three scientific arguments. First, the observed "anthropogenic" increase of concentrations of all long-lived naturally occurring greenhouse gases (carbon dioxide, methane and nitrous oxide); second, the high correlation between CO_2 and CH_4 concentrations on one side and temperature in the lower atmosphere and at the surface on the other for a time period since 160000 years before present; third, projections of massive mean global warming of several degrees within the 21st century by climate models if emissions continue unabated. At that time not really coupled ocean/atmosphere models but atmospheric general circulation models with a mixed layer ocean coupled to them were used, calculating the equilibrium response to a prescribed greenhouse gas increase, that had been derived from simple projections of emissions without climate protection measures.

At the first Conference of the Parties (COP1) to the UNFCCC during end of March/beginning of April 1995 the signatories asked for a legal instrument or a protocol to the convention to be ready at COP3, as it became also clear to politicians that stabilising emissions in industrialised countries at 1990 levels in 2000 was far from being a real climate protection measure. The Kyoto Protocol, accepted by all attending countries (> 150) at COP3 in Kyoto, Japan, asks for greenhouse gas emission reductions, which would, if ratified by 55 countries encompassing at least 55 percent of all CO_2 emissions by all industrialised countries (called Annex 1 countries in the UNFCCC), lead for the first time to a trend reversal in emissions in industrialised countries. Why did this happen? Because the scientific basis got a fourth pillar in the second full assessment report of IPCC (1996): "The balance of evidence suggests a discernible human influence on global climate". The anthropogenic signal had been detected in observations despite the delay of the full signal by decades that is caused largely by the high heat capacity of the ocean.

However, the ratification process became a nearly endless story as some industrialised countries had to realise that reducing emissions needs tough policy-making and interested groups exaggerated scientific uncertainties and climate protection costs. Therefore, many of the flexible mechanisms in the Kyoto Protocol did not get the necessary small print until COP7 in Marrakesh in November 2001, where accounting for carbon sinks in regrowing forests, emissions trading with countries reducing beyond their commitments, joint implementation of measures by industrialised countries and the Clean Development Mechanism (joint implementation between an industrialised and a developing country) got so much attention that reduction commitments can be watered down substantially. On the other side the Kyoto Protocol contains sanctions, e.g. when commitments for the first period measured as an average over 2008 to 2012 are not reached a factor of 1.3 in reductions has to be reached within the second period, that has to be negotiated starting in 2005. As the USA government withdrew in 2001 from the Kyoto process, enactment as a legally binding document needs under all circumstances the ratification by the Russian Federation, which announced to do so soon at the World Summit of Sustainable Development in Johannesburg, South Africa, September 2002.

Why did we finally get the small print for all Kyoto mechanisms? Again through the support of an IPCC report. This time the third assessment (IPCC, 2001a) attributed the rapid global warming of the recent decades predominantly to human action.

Implementing enforcements of the Kyoto Protocol will surely lower the risks caused by weather extremes, but only in the long run. Therefore, it will not lead to decreasing adaptation measures to ongoing climate change for the coming few decades. We live in a period where we have to pay for the absence of measures to slow down the systematic change in atmospheric composition that occurred mainly since the end of the Second World War.

Transformation of the Energy System

The key contribution to rising greenhouse gas levels in the atmosphere originates from the fossil fuel dominated energy supply system in developed and also developing countries that accounts for more than 50 percent of the so-called equivalent CO_2 burden ($CO_2 + CH_4 + N_2O$) to the atmosphere. Therefore, the implementation of Kyoto Protocol commitments will be concentrating in most industrialised countries on measures to decarbonise the energy system both by efficiency increases and energy technology changes. Of all the options available the simplest is a "fuel switch" from lignite or hard coal to natural gas in the power plant and heating sector, decarbonising by up to a factor 2. The option "nuclear energy", which is – disregarding the power supply look-in to a non-renewable energy path – nearly carbon emission free, is none for most developing countries and is already not taken as an option by several highly developed countries. Mainly because of the risks associated with it, especially those related to potential proliferation of nuclear fuel to countries under a dictatorship and to criminals, as well as terrorist attacks against nuclear facilities. The general option "strong energy efficiency increase" is a major one irrespective of the fuel used but the dependence on the fossil path will not be diminished by it, if no long-term activity develops the key option: use of renewable energy sources, i.e. passive solar, solar thermal, photovoltaics, biomass, wind, waves, geothermal.

The sun delivers on average an energy flux of about 170 Wm-2 at the surface. Multiplied by the area of continents and coastal seas this offer amounts to 3.4 • 1016 W, about 3000 times the energy flux needed by humankind at present. Consequently we would not interfere with the natural energy fluxes considerably if this option were chosen. The main problem when transforming our energy system into a second solar era is the lack of internalisation of external costs caused by the fossil fuel related damages to health and environment. A "shining" example for this lack are the German subsidies for hard coal, still amounting to about 3 billion Euros per year. Therefore, the new renewables energy sources must get incentives to leave the "permil ghetto" in most countries. In this context it is not a good policy to only help with market introduction incentives for those renewables like wind energy that are close to being economical, one must also stimulate research and development for renewable like photovoltaics, which will be economical on large scale only in decades from now, however, will allow the least disturbance to natural energy fluxes and atmospheric composition. For example, it would not be wise to invest into massive biomass use for energy, as this would compete for areas needed for food production. Direct solar energy use is the best option in the long run as it does not interfere with the natural climate system above the permitted threshold.

Local Adaptation to and Global Mitigation of Rapid Climate Change

The delayed reaction of the climate system, the late start into climate protection policies (climate protection is here defined only as the dampening of the anthropogenic climate change rate), the adaptation to already ongoing and accelerating global climate change, and the inability to stop the greenhouse gas concentration increase within the next few decades constitute a very challenging situation for mankind. We need to mitigate globally and adapt locally simultaneously but cannot observe global effects within less than several decades. More of our gross national product has to be spent for adaptation to climate change (mostly protecting us from weather extremes), keeping in mind that a major ethical problem will push us ahead: those most vulnerable to global climate change often have not caused it. Therefore, we need to find the financial support for developing countries in their battle to adapt, as it is already foreseen in adaptation fund within the Kyoto Protocol implementation process. A major source for this financing could become the introduction of levies when we use global common goods like air space or the open sea, as recently proposed by the German Global Change Council (WBGU 2002).

References

Caspary HJ, Haeberli W (1999) Klimaänderungen und die steigende Hochwassergefahr. In: Graßl H (ed) Wetterwende: Vision Globaler Klimaschutz. Campus-Verlag, Frankfurt/Main, pp 206-229

Graßl H (2000) Status and improvements of coupled general circulation models-review. Science 288:1991-1997

Haeberli W, Hoelzle M, Maisch M (2001) Glaciers as key indicators of global climate change. In: Lozán J, Graßl H, Hupfer P (eds) Climate of the 21st century: changes and risks. GEO Wissenschaftliche Auswertungen. Hamburg

Houghton JT, Jenkins GJ, Emphraums JS (eds) (1990) Climate change: the IPCC scientific assessment. Cambridge University Press, Cambridge

Houghton JT et al. (eds) (1996) Climate change 1995 – the science of climate change: contribution of working group I to the second assessment report of the Intergovernmental Panel on Climate Change. Cambridge University Press, Cambridge

Houghton JT et al. (eds) (2001) Climate change 2001: the scientific basis. Working group I contribution to the Third Assessment Report of the IPCC. Cambridge University Press, Cambridge

IPCC (Intergovernmental Panel on Climate Change) (ed) (2001): Klimaänderung 2001–Zusammenfassung für politische Entscheidungsträger. ProClim, Bern

Palmer TN, Räisänen J (2001) Quantifying risk in a changing climate. CLIVAR Newsletter 6(3):3-4

Semenov VA, Bengtsson L (2002) Secular trends in daily precipitation characteristics: greenhouse gas simulation with a coupled AOGCM. Climate Dynamics 19:123-140

WBGU (German Advisory Council on Global Change) (ed) (2002) Charging the use of global commons. WBGU, Berlin

2.2 Assessing Human Vulnerability to Global Climatic Change

Mike Brklacich[1], Hans-Georg Bohle[2]

[1]Global Environmental Change and Human Security Project and Department of Geography and Environmental Studies, Carleton University, Ottawa
[2]Department of Geography, University of Bonn, Meckenheimer Allee 166, 53115 Bonn

Introduction

Concerns about global climatic change have shifted substantially over the past two decades. Scientific evidence has confirmed human activities have altered the earth's atmosphere to such an extent that wide scale climatic changes are anticipated. One of the consequences stemming from these scientific advances has been a growth in concerns related to the societal consequences of climatic change. This paper traces the development of climatic change – society research and argues there is an urgent need to re-orient this research field in order to incorporate assessments of human vulnerability.

Two generations of Climate Change Assessments

First generation of Assessments: Impact of Climatic Change

Much of the early research into climatic change – societal relationships has been based upon the "impacts" research framework illustrated in Figure 2.2.1. For the purposes of this paper, agricultural examples are employed to assist with explaining the impacts framework but the basic approach has been applied to several other economic sectors and human activities.

Impact studies usually commence with the specification of several climatic change scenarios. While there is little doubt that human activities are and will continue to alter climate systems on a global scale, considerable uncertainty remains about the pace and magnitude of climatic change. Scenarios are regularly deployed to capture this uncertainty and many impact studies will employ "best and worst case" scenarios. Climatic change scenarios are often derived from the

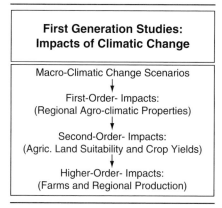

Fig. 2.2.1 First Generation Studies:
Impacts of Climatic Change

outputs of General Circulation Models (GCMs) but spatial and historical analogues have also been used to develop scenarios for future climates. The art and science of developing comprehensive scenarios has been developing rapidly and most contemporary scenarios for climatic change are founded upon sets of assumptions regarding long-term prospects for economic development, governance, population and demography and so on.

These climatic change estimates are usually provided at a scale which are too broad for impact assessments and hence the second step involves interpolating these macro scenarios for climatic change into regional climatic change scenarios. These first-order impacts involve down-sizing the scale of the macro-climatic change scenarios as well as converting basic climate data such as change in mean daytime temperature and daily precipitation into parameters that are useful within the context of the specific study. For example, basic climate data would be converted into the start and end of the frost-free season for climatic change – agricultural impact studies.

Second-order impacts typically estimate the impacts of climatic change on primary economic activities. For climatic change – agricultural studies, agricultural land suitability and crop yield models have been used extensively to estimate the potential effects of estimated changes to climate on where particular crops might be grown and on crop yields. Higher-order impacts typically employ the outputs of second-order assessments to gauge the effects of climate change on agricultural production and profitability at farm through regional levels.

Second generation of Assessments: Responses to Climatic Change

Second generation assessments recognised that socio-economic systems are dynamic and regularly respond to external stimuli. Second generation assessments employ the same basic research framework as first generations assessments but extend the analyses to include responses to impacts emanating from global climatic change (Figure 2.2.2). Overall, second generation studies add feedback loops to the impacts research framework and responses can be put into three broad categories: non-response, mitigation and adaptation. In all cases, the human activities associated with the various responses can alter future climatic changes.

These second generation studies acknowledge that impacts will not always prompt a response and hence the inclusion of the non-response category. Non-response could occur for several reasons including impacts not representing a threat to the socio-economic sector in question, there being no known technology to overcome negative impacts, and economic, political and technological constraints preventing a response (e.g. lack of knowledge of response options, insufficient capital to adapt, weak institutions diminish the likelihood of a response).

The mitigation and adaptation options incorporate changes in human behaviour as well as climate however the basic tenants underpinning these two response options are fundamentally different. Mitigation assessments are based upon preventing, or at least reducing the pace and rate of global climatic change. Mitigation options can include approaches which would reduce greenhouse gas emission levels at the source (i.e. improving fuel-use efficiency and conversion to alternative energy sources which are not reliant on fossil fuels) as well as carbon sequestering which assists with reducing atmospheric CO_2 levels (e.g. reforestation and rebuilding of soil carbon stocks).

This preventative approach is contrasted by adaptation which investigates how changes in human behaviour could either offset negative impacts stemming from climatic change or, in the case of positive impacts, allow an economic sector to capitalise on these new opportunities. In other words, it is assumed that climatic change has and will continue to occur and the primary driving force behind this response option involves investigating and implementing adjustment strategies.

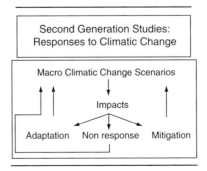

Fig. 2.2.2 Second Generation Studies: Responses to Climatic Change

Attributes of first and second generation Assessments

Several attributes are common to both first and second generation climatic change assessments. Some of their key attributes include, the assessments:

- are relative to a single external stimuli (i.e. climatic change)
- assume climatic change will occur slowly and incrementally,
- are conducted at one spatial scale and do not consider cross-scale impacts,
- are triggered by climatic change and therefore there is only limited consideration of socio-political context in which climatic change occurs, and
- consider human vulnerability as a residual of climatic change impacts.

The cumulative effect of these common attributes is that the first and second generation assessments are well-suited to isolating the effects of single stressors (e.g. impact of climatic change on crop yields) and on a single sector (e.g. agriculture, forestry) as well as gauging the technical efficiency of a specified response option (e.g. effectiveness of irrigation and/or alternative cultivars as a means to offset crop yield reductions stemming from drier climates). Recent advances in estimating climatic change at various spatial scales, improvements in impact assessment methods and a more thorough understanding of response options continue to improve the accuracy of assessments based upon these first two generations of climatic change assessments. Nevertheless, there is a growing demand to broaden the basis for climatic change assessments and to explicitly engage with the concepts relating to risk and vulnerability. The first two generations of climate change research were however never designed to treat societal concerns in a holistic fashion and are not well-equipped to explicitly consider human vulnerability to climatic change.

Towards a third generation Climatic Change Assessments: A Human Vulnerability Focus

Foundations of Human Vulnerability Concepts

Even though few climatic change studies have explicitly assessed human vulnerability, it is not a new concept. For example, natural hazards research usually defines vulnerability as the potential for loss and distinguishes between biophysical and social vulnerability (Cutter 1996). Biophysical or inherent vulnerability is associated with the attributes or characteristics (e.g. magnitude, duration, frequency) associated with a naturally occurring event such as an earthquake, flood or cyclone. The impacts on human activities and human well-being of natural phenomena are very seldom uniform as the event is mediated through a range of social, economic and political conditions. Hazards research recognises impacts on human activities are a function of both the events characteristics and broader societal conditions, and hence the study and assessment of natural hazards focuses on the interaction between natural events and human systems and not the natural system in isolation (Mustafa 1998).

Early work into the mitigation of human impacts stemming from natural hazards usually focused on the natural phenomena and concentrated on technocratic and engineering solutions such as levees to reduce flood risk and irrigation systems to reduce drought risks. This sort of solutions has without doubt decreased human injury and loss of life. This reliance on technology has, however, been criticised for at least two reasons. First, they often encourage intensive human development in hazard prone areas which can lead to massive losses in investment and human lives when the engineering measure fails. Floods in the Mississippi and Saguenay River Valleys in the USA and Canada respectively during the 1990s serve as vivid examples of the consequences of an over reliance on engineering approaches which also shifts the focus away from the underlying social-economic-political factors that play crucial roles in determining the human impacts of extreme events. Another concern is that the high capital cost associated with implementing engineering solutions is often beyond the economic grasp of poorer countries and communities. Concerns about the equity of many engineering solutions have been raised and their tendency to reinforce divides between "have" and "have-not" regions has also been brought into question.

Behavioralist approaches have expanded the field of hazards research beyond technocratic and engineering perspectives to include approaches which address questions relating to how individuals and institutions perceive hazards (Burton et al. 1993) as well as approaches which explicitly focus on the social and political causality of human vulnerability to hazards (Blaikie et al. 1994, Watts and Bohle 1993, Davis 2001). This latter approach has led to the development of comprehensive notions of human vulnerability to hazards and related phenomena such as hunger and famine. In this broadened context, vulnerability is not viewed as a result of an extreme event but rather as a societal characteristic or property which is present in a social system prior to the onset of the extreme event. Earlier work by Watts and Bohle (1993) explored the social space of vulnerability and revealed how famine was attenuated by access to resources, power relations and class relations. In this context, human vulnerability was developed as a three-dimensional concept which focuses attention explicitly on the underlying socio-political causes which triggered famine rather than studying an extreme natural phenomena such as a seasonal drought. Benefits stemming from this approach include insight into how human vulnerability varies amongst subpopulations within a region as well as an improved understanding of whether the primary cause of a famine is attributable to inadequate access to agricultural resources such as land and water or weak institutions or concentration of power or some combination of these factors. This basic approach has been developed further and applied to several environmental hazards (Adger 1999, Cutter 2000, Mustafa 1998) and most recently to non-conventional threats to human security (Watts 2002).

Moving from Hazards to Climate Change

Natural hazards and famine-related research typically consider relatively narrow time frames over which societal changes would be minimal and therefore it is legitimate to assume societal change as exogenous to specific case studies. Climatic change is however expected to occur over a longer period and hence it is inappropriate and misleading to assume stable social systems (Clark 1985, Wilbanks and Kates 2001). This does not suggest that the basic tenets which underpin human vulnerability concepts as developed within a natural hazards context are unworkable in a climatic change context but it does suggest that a dynamic element will need to be added to these essentially static frameworks in order to capture changes in both climate and social systems. For example, economic globalisation has become a set of pervasive forces which are contributing to fundamental and wide sweeping changes in economic, social and political systems. In addition to further concentration of wealth and increase in competition amongst multi-national firms, economic globalisation is challenging and changing relationships between nation-states and the private sector. As a result, the structure and role of both formal and informal institutions are being re-defined which in turn are altering human vulnerability to environmental threats and environmental change. It is in this context that human vulnerability can be magnified by the synergistic effects stemming from a more variable climate as well as the consequences of economic globalisation (O'Brien and Leichenko 2000).

Figure 2.2.3 represents one approach to capturing the combined effects of both climatic and societal change on human vulnerability. This framework defines human vulnerability in the context of exposure to climatic change and the capacity to cope with and recover from climatic change. Exposure is sensitive to both climatic and societal changes. For example, a future climate characterised by more frequent and intense storms would result in greater exposure only if formal and informal institutions which currently insulate society from climatic events were not able to adequately protect society from the expected increases in more severe weather. Overall, assessments of exposure to climatic change need to consider both climatic and societal changes.

Coping and recovery capacity is also cast in the context of climatic and societal change. More frequent storms could contribute to less than full economic recovery between storms and these cumulative effects could stretch the coping capacity of a particular community beyond its current limit. From a societal change perspective, a weakened economy could make recovery more difficult and thereby reduce the capacity to cope with the next climatic extreme.

The approach outline in Figure 2.2.3 develops vulnerability as a dynamic concept which is sensitive to both changes in exposure and coping capacity. It does not view vulnerability as a residual of climatic change impact but rather it positions human vulnerability as precursor to impacts. If a community possesses sufficient coping capacity, then it would follow that an altered climate would not necessarily result in economic and social disruptions. Similarly, it may be possible

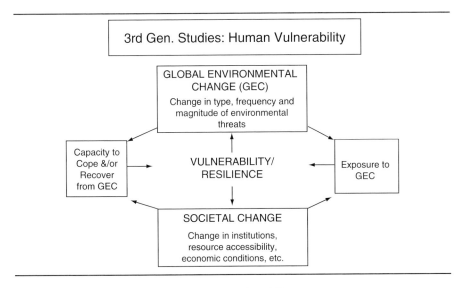

Fig. 2.2.3 Third Generation Studies: Human Vulnerability

to bolster the capacity of a community to cope and thereby offset potential negative impacts associated with climatic change. Overall, the approach provides a foundation for assessing the relative merits of mitigating climatic change compared to bolstering adaptation capacity.

Further Development of Vulnerability Concepts

Much work has been done on the vulnerability topic, but little has so far been developed in terms of concepts and theories that might help to more systematically explore environmental and social vulnerabilities. Figure 2.2.4 is a simplistic first attempt to pull together, by no means exhaustively, some theoretical debates relevant to environmental and social vulnerabilities. The model includes new strands of thinking in social and environmental theory. It builds upon the Environmental Entitlements Program, which has been developed by the Institute of Development Studies in Sussex and has since been taken up by various development agencies, focuses on community-based sustainable development (see e.g. Leach, Mearns and Scoones 1997). The programme focuses upon who regulates access to and command over scarce, contested environmental goods and services, and in this context, vulnerability rises as access to environmental goods and services is impeded. The programme draws upon an array of theoretical considerations, including social theory, environmental theory, entitlements theory and conflict theory. The issue of institutions is also widely discussed, because institutions play a central role in mediating the relationships between environment and society, between land and land managers.

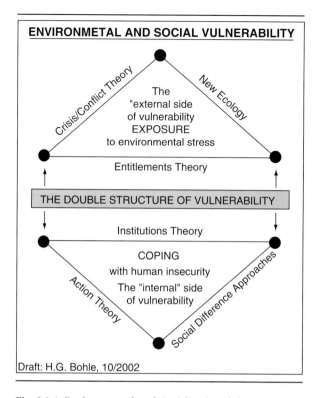

Fig. 2.2.4 Environmental and Social Vulnerability

Figure 2.2.4 emphasises the double structure of vulnerability with the "coping component" focusing on social vulnerability and the "exposure component" concentrating on environmental vulnerability. Social difference theory is taken up to show that gender, caste, wealth, age, origins and other aspects of social identity divide and crosscut community boundaries. This study emphasises how diverse and often conflicting values and resource priorities in degraded environments – rather than shared beliefs and interests – pervade social life and may be struggled and "bargained" over (e.g. Carney and Watts 1997). Equally, social difference theory draws attention to power as a pervasive feature of land management (e.g. Gaventa 1995).

Another strand of social theory, action theory, argues that land managing communities cannot be treated as static, undifferentiated wholes, since they are composed of active people and groups. The behaviour of social actors is not driven automatically and unconsciously by structures of environmental change and land degradation; rather they actively monitor, interpret and shape the environment around them (e.g. Long and Long 1992). Action theory has also grappled with the links between agency and structure, emphasising how structures, rules and norms can emerge (and only exist) as products of people's land use practices and actions, both intended and unintended. These structural forms subsequently shape people's actions, not by strict determination but by providing flexible orientation points, which either constrain or enable what is possible in community-based sustainable development under conditions of global environmental change (e.g. Long and van der Ploeg 1994).

The so-called "new ecology" approach (e.g. Botkin 1990) stresses that spatial and temporal variability in land degradation which contribute to dynamic, non-equilibrial processes of land use and histories of disturbance events. Just as communities cannot be treated as static or undifferentiated, made up as they are of active land managers, the environment equally needs to be disaggregated into its constituent parts (soil, water, vegetation, rocks, climate, relief), and viewed dynamically.

Entitlements theory (e.g. Sen 1981; for a new critical discussion, see Watts 2002) is employed to explain how the consequences of land degradation and access to and control over land are also socially differentiated. Entitlements refer to legitimate effective command over alternative commodity bundles, i.e. land, water and workforce. More specifically, environmental entitlements refer to alternative sets of utilities derived from environmental goods such as land and water over which social actors have legitimate effective command, the management of which are instrumental in achieving well-being.

Violent conflict, frequently fought along ethnic cleavages, is certainly one of the most dramatic and increasingly widespread threats to human security. From the perspective of conflict theories based on rational choice theory, "ethnic" conflicts have been conceptualised as stemming from "structural antagonisms" which may lead to mobilising "cultural resources". This is why violent conflict can become so highly emotional and extremely violent (Esser 1998). Others view violent conflict also as "structural conflicts", but explain the mobilisation of violence as a reaction to a new "global disorder" (IDS Bulletin 2001). Still another approach has been propagated by Wimmer (2002) who explains widespread violent conflict as an outcome of nationalist exclusion. This process is most acute in the process of nation-building, when groups in a privileged relationship to the modernising state may enjoy democratic participation, equality before the law and self-determination, while minority groups ("ethnic" minorities) are denied these privileges, which frequently leads to brutal civil wars. The key point here is that crises and conflict can destabilise societies and thereby increase exposure to environmental stress.

Last but not least environmental entitlements concepts also draw on institutions theory (North 1990). Institutions play a central role in mediating the relationships between environment and society, between land and agriculturists. Many analysts have described institutions either as "rules" or as "organisations" and more recently, institutions have rather been conceptualised as "rules in action" or "regularised practices". The central question that emerged is asking which combination of institutions makes the most difference to resource access and control for a set of social actors, or for the dynamics of resources use and management surrounding particular valued elements of the landscape. Community-based sustainable development is always shaped by institutions that are embedded in structures of power relations, that are made and remade, and that are highly contested through people's practice.

Conclusion

This paper has traced the development of climatic change studies and argued there is an urgent need to fully incorporate human vulnerability concepts into this research field. The roots of research into human vulnerability to environmental stresses can be traced back to natural hazards research and over the past decade there has been a growing recognition that human vulnerability has both an external component which determines exposure to environmental stress and an internal component which examines human capacity to cope. The development of vulnerability concepts has to a large extent occurred outside the climatic change community. There appears to be a solid foundation better for incorporating human vulnerability within climatic change research and the vulnerability frameworks illustrated in this paper provide a basis for re-orienting climatic change studies. There is however an urgent need to build upon these vulnerability frameworks and develop a set of placed-based research programs to test the strengths of these frameworks within a climatic change setting.

Acknowledgements

This research has benefited from support from the Social Science and Humanities Research Council of Canada.

References

Adger W (1999) Social vulnerability to climate change and extremes in coastal Vietnam. World Development 27(2):249-269

Blaikie P (1994) At risk: natural hazards, people's vulnerability and disasters. Routledge, London

Bohle HG (2001) Vulnerability and criticality: perspectives from social geography. IHDP Update 2:1-5

Burton I, Kates R, White G (1993) The environment as hazard. Guilford, New York

Carney J, Watts M (1991) Manufacturing dissent: work, gender and the politics of meaning in a peasant society. Africa 60(2):207-241

Clark W (1985) Scales of climatic change. Climatic Change 7(1):5-27

Cutter S (1996) Vulnerability to environmental hazards. Progress in Human Geography 20(4):529-539

Cutter S, Mitchell J, Scott M (2000) Revealing vulnerability of people and places: A case study of Georgetown, South Carolina. Annals of the Association of American Geographers 90(4):713-737

Davis M (2001) Late Victorian holocausts: El Nino and the making of the Third World. Verso, London

Esser H (1996) Die Mobilisierung ethnischer Konflikte. In: Bonde KJ (ed) Migration, Ethnizität, Konflikt: Systemfragen und Fallstudien. Universitätsverlag Rasch, Osnabrück, pp 63-87

Gaventa J (1995) Citizen knowledge, citizen competence and democracy building. The Good Society 5(3):28-35

Institute of Development Studies (2001) Structural conflict in the new global disorder. IDS-Bulletin 32(2)

Long N, Long A (eds) (1992) Battlefields of knowledge: the interlocking of theory and practice in social research and development. Routledge, London

Long N, van der Ploeg JD (1989) Demythologizing planned intervention: an actor's perspective. Sociologia Ruralis 29(3/4):227-249

Mustafa D (1998) Structural causes of vulnerability to flood hazard in Pakistan. Economic Geography 74(3):289-305

North DC (1990) Institutions, institutional change and economic performance. Cambridge University Press, Cambridge

O'Brien K, Leichenko R (2000) Double exposure: assessing impacts of climatic change within the context of economic globalization. Global Environmental Change 10(3):221-232

Sen A (1981) Poverty and famines. An essay on entitlement and deprivation. Clarendon Press, Oxford

Watts M (2002) Hours of darkness: vulnerability, security and globalization. Geographica Helvetica 57(1):5-18

Watts M, Bohle H (1993) The space of vulnerability: the causal structure of hunger and famine. Progress in Human Geography 17:43-67

Wilbanks T, Kates R (1999) Global change in local places: how scale matters. Climatic Change 43:601-628

Wimmer A (2002) Nationalist exclusion and ethnic conflict. Shadows of modernity. Cambridge University Press, Cambridge

2.3 Fire Ecology of the Recent Anthropocene

Johann G. Goldammer

Max-Planck-Institute for Chemistry, Joh.-Joachim-Becher-Weg 27, 55128 Mainz

Preamble

Fire has the capacity to make or break sustainable environments. Today some places suffer from too much fire, some from too little or the wrong kind, but everywhere fire disasters appear to be increasing in both severity and damages, with serious threats to public health, economic well being, and ecological values (Pyne 2001). Thus, fire ecology in the recent Anthropocene has evolved to a science of the biosphere with strong multi-disciplinary interconnectedness between natural sciences – notably ecology, biogeochemistry, atmospheric chemistry, meteorology and climatology – and the humanities – anthropology, cultural history, sociology, and political sciences (Goldammer 1993, Goldammer and Crutzen 1993).

Introduction

Fire is a prominent disturbance factor in most vegetation zones across the world. In many ecosystems it is an essential and ecologically significant force -organising physical and biological attributes, shaping landscape patterns and diversity, and influencing energy flows and biogeochemical cycles, particularly the global carbon cycle. In some ecosystems, however, fire is an uncommon or even unnatural process that severely damages the vegetation and can lead to long-term degradation. Such ecosystems, particularly in the tropics, are becoming increasingly vulnerable to fire due to growing population, economic and land-use pressures. Moreover, the use of fire as a land-management tool is deeply embedded in the culture and traditions of many societies, particularly in the developing world. Given the rapidly changing social, economic and environmental conditions occurring in developing countries, marked changes in fire regimes can be expected, with uncertain local, regional, and global consequences. Even in regions where fire is natural (e.g. the boreal zone), more frequent severe fire weather conditions have created recurrent major fire problems in recent years. The incidence of extreme wildfire events is also increasing elsewhere the world, with adverse impacts on economies, livelihoods, and human health and safety that are comparable to those associated with other natural disasters such as earthquakes, floods, droughts and volcanic eruptions. Despite the prominence of these events, current estimates of the extent and impact of vegetation fires globally are far from complete. Several hundred million hectares of forest and other vegetation types burn annually throughout the world, but most of these fires are not monitored or documented. Informed policy and decision-making clearly requires timely quantification of fire activity and its impacts nationally, regionally and globally. Such information is currently largely unavailable.

The primary concerns of policy makers focus on questions about the regional and global impacts of excessive and uncontrolled burning, broad-scale trends over time, and the options for instituting protocols that will lead to greater control. Other key questions involve determining in what circumstances fires pose a sufficiently serious problem to require action; what factors govern the incidence and impacts of fires in such cases; and what might be the relative costs and benefits of different options for reducing adverse impacts? The elaborations in this paper reflect the rationale and the scope of work of the Working Group on Wildland Fire of the United Nations International Strategy for Disaster Reduction (ISDR) (Working Group on Wildland Fire 2002) with its associated international wildland fire research programmes of the International Geosphere-Biosphere Programme (IGBP) and other research consortia.

Extent, Impacts and Significance of Wildland Fire at Global Scale

Global Wildland Fire Assessments

The Forest Resources Assessment 2000 conducted by the Food and Agriculture Organisation of the United Nations (FAO) provided an opportunity to review the global effects of fires on forests as a part of the forest assessment that is undertaken every ten years. The Global Forest Fire Assessment 1990-2000 (FAO 2001), prepared in cooperation with the Global Fire Monitoring Center (GFMC), revealed strengths and weaknesses associated with sustaining the health and productivity of the world's forests when threatened by drought, wildfires and an increasing demand for natural resources.

Much of the materials for the Global Forest Fire Assessment had been taken from the Global Fire Monitoring Center (GFMC) database or recruited through the GFMC global fire network. Beginning in the late 1980s the GFMC started to systematically collect worldwide fire information (statistical data, narratives) and published these materials in FAO-ECE International Forest Fire News (IFFN) and on the GFMC website (GFMC 2002). Together with the ECE fire database the GFMC fire statistical information represents a unique source that includes also information on fires occurring in other vegetation types, e.g., savannah fires and agricultural burning.

However, format and completeness of wildland fire statistics collected are not consistent. Statistical datasets providing number of fires and area burned do not meet the demand for information required to assess the environmental and economic consequences of wildland fires. For instance, currently used formats for fire statistics collection do not include parameters that would allow to conclude on economic damages or impacts of emissions on the atmosphere or human health. Considering the complexity of pathways of regeneration of vegetation after fire, including the cumulative impacts of anthropogenic and environmental stresses, it is not possible at the moment to conclude from existing statistical data whether long-term changes can be expected in terms of site degradation and of reduction of carrying capacity of fire-affected sites. Thus, a new system for fire data collection that would meet the requirements of different users is urgently needed.

The recently published Global Burned Area Product 2000 derived from a spaceborne sensing system (SPOT Vegetation) is a first and important step towards obtaining prototype base data on the extent of global wildland fires for the year 2000 (JRC 2002). The dataset indicates a size of ca. 351 million hectares affected by fire in the year 2000 globally. Looking at individual countries the datasets differ considerably from the numbers provided by responsible national agencies. This fact that can be explained easily because statistical datasets of national agencies in many countries primarily include fire incidences in managed forests; only in a few number of countries fire statistics cover non-forest ecosystems. Furthermore most countries do not have in place appropriate means to survey wildland fire occurrence impacts. On the other hand the area burned, as derived from the spaceborne instrument, does also not provide information on environmental, economic or humanitarian impacts of fire. An appropriate interpretation of satellite-generated fire information requires additional information layers, particularly on ecosystem vulnerability and recovery potential; these are not yet available at global level.

Summarising this state-of-the-knowledge review for the last decade of the 20th century the Global Forest Fire Assessment came to this conclusion:

- Wildfires during drought years continue to cause serious impacts to natural resources, public health, transportation, navigation and air quality over large areas. Tropical rain forests and cloud forests that typically do not burn on a large scale were devastated by wildfires during the 1990s.
- Many countries, and regions, have a well-developed system for documenting, reporting and evaluating wildfire statistics in a systematic manner. However, many fire statistics do not provide sufficient information on the damaging and beneficial effects of wildland fires.
- Satellite systems have been used effectively to map active fires and burned areas, especially in remote areas where other damage assessment capabilities are not available.
- Some countries still do not have a system in place to annually report number of fires and area burned in a well-maintained database, often because other issues like food security and poverty are more pressing.

- Even those countries supporting highly financed fire management organisations are not exempt from the ravages of wildfires in drought years. When wildland fuels have accumulated to high levels, no amount of firefighting resources can make much of a difference until the weather moderates (as observed in the United States in the 2000 fire season).
- Uncontrolled use of fire for forest conversion, agricultural and pastoral purposes continues to cause a serious loss of forest resources, especially in tropical areas.
- Some countries are beginning to realise that inter-sectoral coordination of land use policies and practices is an essential element in reducing wildfire losses.
- Examples exist where sustainable land use practices and the participation of local communities in integrated forest fire management systems are being employed to reduce resource losses from wildfires.
- In some countries, volunteer rural fire brigades are successful in responding quickly and efficiently to wildfires within their home range; and residents are taking more responsibility to ensure that homes will survive wildfires.
- Although prescribed burning is being used in many countries to reduce wild-fire hazards and achieve resource benefits, other countries have prohibitions against the use of prescribed fire.
- Fire ecology principles and fire regime classification systems are being used effectively as an integral part of resource management and fire management planning.
- Fire research scientists have been conducting cooperative research projects on a global scale to improve understanding of fire behaviour, fire effects, fire emissions, climate change and public health.
- Numerous examples were present in the 1990s of unprecedented levels of inter-sectoral and international cooperation in helping to lessen the impact of wildfires on people, property and natural resources.
- Institutions like the Global Fire Monitoring Center have been instrumental in bringing the world's fire situation to the attention of a global audience via the Internet.

In reviewing the global fire situation, it is apparent that a continued emphasis on the emergency response side of the wildfire problem will only result in future large and damaging fires. The way out of the emergency response dilemma is to couple emergency preparedness and response programmes with more sustainable land use policies and practices. Only when sustainable land use practices and emergency preparedness measures complement each other do long-term natural resource benefits accrue for society.

In the following sections an updated review is provided on the changing fires regimes in the most important vegetation zones, the northern boreal forest, temperate forests, tropical rainforest and tropical / subtropical savannah regions.

Fire Situation in the Boreal Forests

The global boreal forest zone, covering approximately 12 million square kilometres, stretches in two broad transcontinental bands across Eurasia and North

America, with two-thirds in Russia and Scandinavia and the remainder in Canada and Alaska. With extensive tracts of coniferous forest that have adapted to, and become dependent upon, periodic fire for their physiognomy and sustainable existence, and that provide a vital natural and economic resource for northern circumpolar countries, boreal forests are estimated to contain ~37% of the world's terrestrial carbon. These forests have become increasingly accessible to human activities, including natural resource exploitation and recreation, over the past century, with the export value of forest products from global boreal forests accounting for 47% of the world total.

The largest boreal forest fires are extremely high-intensity events, with very fast spread rates and high levels of fuel consumption, particularly in the deep organic forest floor layer. High intensity levels are often sustained over long burning periods, creating towering convection columns that can reach the upper troposphere and lower stratosphere, making long range smoke transport common. In addition, the area burned annually by boreal fires is highly episodic, often varying by an order of magnitude between years.

For many reasons boreal forests and boreal fires have increased in significance in a wide range of global change science issues in recent years. Climate change is foremost among these issues, and the impacts of climate change are expected to be most significant at northern latitudes. Forest fires can be expected to increase sharply in both incidence and severity if climate change projections for the boreal zone prove accurate, acting as a catalyst to a wide range of processes controlling boreal forest carbon storage, causing shifting vegetation, altering the age class structure towards younger stands, and resulting in a direct loss of terrestrial carbon to the atmosphere.

Over the past two decades forest fires in boreal North America (Canada and Alaska) have burned an average of 3 million hectares annually. While sophisticated and well-funded fire management programs in North America suppress the vast majority of fires while small, the 3% of the fires that grow larger than 200 hectares in size account for 97% of the total area burned. There is little likelihood of improving suppression effectiveness since the law of diminishing returns applies here – the 3% of fires that grow larger do so because they occur under extreme fire danger conditions and/or in such numbers that suppression resources are overwhelmed, and applying more funding would have no effect. In addition, Canada and Alaska have vast northern areas which are largely unpopulated and with no merchantable timber, and where fires are monitored but suppression is unwarranted and not practiced, allowing natural fire where possible. Combining the current high levels of fire activity across the North American boreal zone, with restricted suppression effectiveness and a recognition of the need for natural fire, it is all but certain that climate change will greatly exacerbate the situation, and the only option will be adaptation to increasing fire regimes.

The fire problem in Eurasia's boreal forests – mainly in the Russian Federation – is similar in some ways to the North American situation, but there are significant differences (Goldammer and Furyaev 1996, Goldammer and Stocks 2000,

Shvidenko and Goldammer 2001). The Eurasian boreal zone is close to twice as large as its North American counterpart, stretching across eleven time zones, with a wide diversity of forest types, growing conditions, structure and productivity, and anthropogenic impacts that define different types of fires and their impacts. For example, surface fires are much more common in the Eurasian boreal zone than in North America where stand-replacing crown fires dominate the area burned, and many Russian tree species have adapted to low-intensity fires. While Canada has a strong continental climate in the boreal region of west-central Canada, where most large fire activity occurs, Eurasia has much stronger continentality over a much larger land base. The climate of the major land area of the Russian Federation is characterised by low annual precipitation and/or frequent droughts during the fire season, and more extreme fire danger conditions over a much larger area. In addition, large areas of the Eurasian boreal zone are not protected or monitored, so fire activity is not recorded for these regions.

Official Russian fire statistics for the past five decades typically report annual areas burned between 0.5 and 1.5 million hectares, with very little inter-annual variability. In fact, as Russian fire managers agree, these numbers are a gross underestimation of the actual extent of boreal fire in Russia. There are tow reasons for this: a lack of monitoring/documentation of fires occurring in vast unprotected regions of Siberia, and the fact that the Russian reporting structure emphasised under-reporting of actual area burned to reward the fire suppression organisation. In recent years, with the advent of international satellite coverage of Russian fires in collaboration with Russian fire scientists, more realistic area burned estimates are being generated. For example, during the 2002 fire season satellite imagery revealed that about 12 million hectares of forest and non-forest land had been affected by fire in Russia. Official sources report 1.2 million hectares forest land and 0.5 million ha non-forest land burned in the protected area of 690 million hectares (Goldammer 2003, Davidenko and Eritsov 2003, Sukhinin et al. 2003).

In recent years there has been an increase in the occurrence of wildfires under extreme drought conditions (e.g. the Trans-Baikal Region in 1987 and the Far East in 1998), and the severity of these fires will greatly disturb natural recovery cycles. There is growing concern that fires on permafrost sites will lead to the degradation or disappearance of eastern Siberian larch forests. Russia has close to 65% of the global boreal forest, an economically and ecologically important area that represents the largest undeveloped forested area of the globe. With the vast quantities of carbon stored in these forests, Russian forests play a critical role in the global climate system and global carbon cycling. Projected increases in the severity of the continental climate in Russia suggest longer fire seasons and much higher levels of fire danger. The inevitable result will be an increase in the number, size, and severity of boreal fires, with huge impacts on the global carbon cycle and the Russian economy.

At the same time, recent political and economic changes in Russia have led to an extreme reduction in their fire suppression capability, with the State Forest Service showing a huge debt, and the outlook for future funding increases bleak.

As a result, *Avialesookhrana*, the aerial fire protection division of the Forest Service, has been forced to reduce aerial fire detection and suppression levels, with operational aircraft flying hours and firefighter numbers now less than 50% of 1980s levels. Consequently, the occurrence of larger fires is increasing, and the areas being burned annually are unprecedented in recent memory.

Increasing fire risks in the boreal forests of Eurasia are a major threat to the global carbon budget, and this requires significant national and international attention (Kajii et al. 2003). Management and protection of these vital resources should not be given solely to the private sector or delegated to regional levels. The establishment and strengthening of a central institution to protect forests must be a priority supported by Russia and the international community.

Fire Situation in Temperate Forests

Temperate North America: The United States

In recent years fires have been increasing in number and severity across the temperate zone, with significant fire events becoming more common in the United States, the Mediterranean Basin, and Mongolia for example.

Over most of the past century, the United States has made a huge investment in wildland fire management, developing a world-quality fire suppression capability. As the 1900s progressed, the United States became increasingly effective in excluding fire from much of the landscape. Despite numerous human-caused and lightning fires the area burned was greatly reduced from the early 1990s. However, the price for successful fire exclusion has proven to be twofold. The most obvious price was the huge cost of developing and maintaining fire management organisations that were increasingly requiring higher budgets to keep fire losses at an acceptable level. The second cost, hidden for decades, is now apparent, as the policy emphasis on fire exclusion has led to the build-up of unnatural accumulations of fuels within fire-dependent ecosystems, with the result that recent fires burn with greater intensity and have proven much more difficult to control. Despite extensive cooperation from other countries, and huge budget expenditures, intense droughts in 2000 and 2002 contributed to widespread wildfires in the western United States that burned between 2.5 and 3 million hectares in each year. Losses from these fires now include substantial destruction of homes, as the trend towards living in fire-prone environments grows, and the wildland urban interface is expanding.

The Mediterranean Basin and the Balkans

Within the Mediterranean Basin fire is the most important natural threat to forests and wooded areas. Mediterranean countries have a relatively long dry season, lasting between one and three months on the French and Italian coasts in the north of the Mediterranean, and more than seven months on the Libyan and Egyptian coasts to the south. Currently, approximately 50,000 fires burn throughout the Mediterranean Basin, and burn over an annual average of 600,000 hectares, both statistics are at a level twice that of the 1970s. In countries where data is continuous since the 1950s, fire occurrence and area burned levels have shown large increases since the 1970s in Spain, Italy, and Greece. Human-caused fires dominate in the Mediterranean Basin, with only 1-5% of fires caused by lightning. Arson fires are also quite prevalent.

Paradoxically, the fundamental cause of an increasing vulnerability of the vegetation of the Southern European countries bordering the Mediterranean Basin is linked to increased standards of living among the local populations (Alexandrian et al. 2000). Far-reaching social and economic changes in Western Europe have led to a transfer of population from the countryside to the cities, a considerable deceleration of the demographic growth, an abandonment of arable lands and a disinterest in the forest resource as a source of energy. This has resulted in the expansion of wooded areas, erosion of the financial value of the wooded lands, a loss of inhabitants with a sense of responsibility for the forest and, what is important, an increase in the amount of fuel.

The demographic, socio-economic and political changes in many countries of Southeast Europe and the neighbouring nations on the Balkan have resulted in an increase of wildfire occurrence, destabilisation of fire management capabilities and increased vulnerability of ecosystems and human populations. The main reasons for this development include the transition from centrally planned to market economies, national to regional conflicts, creation of new nations, involving political tensions and war, land-use changes, and regional climate change towards increase of extreme drought occurrence. New solutions are required to address the increasing fire threat on the Balkans. Regional cooperation in the Mediterranean Region and the Balkans must address the underlying causes of changing fire regimes and a more economic trans-national use of fire management resources (Goldammer 2001b).

Transition from Temperate Steppes to Boreal Forests
With an area of 1,565,000 square kilometres and a population of 2.3 million, Mongolia is one of the least populated countries in the world, yet significant fire problems exist there. With an extremely continental climate, poor soil fertility and a lack of surface water, wildfires in Mongolia have become a major factor in determining the spatial and temporal dynamics of forest ecosystems. Of a total of 17 million hectares of forests, 4 million hectares are estimated to be disturbed, primarily by fire (95%) and logging (5%). Forests are declining in Mongolia as continual degradation by wildfire turns former forests into steppe vegetation. The highest fire hazard occurs in the submontane coniferous forests of eastern Mongolia, where highly flammable fuels, long droughts, and economic activity are most common. In recent years fire activity in Mongolia has increased significantly, due to economic activity on lands once highly controlled or restricted. In Mongolia, only 50-60 forest fires and 80-100 steppe fires occur annually on average. The underlying causes of the majority of wildfires are due to the fact that urban people are seeking for livelihood in the forests because of the collapse of the indus-

trial sector. During the 1996 to 1998 period, when winter and spring seasons were particularly dry, Mongolia experienced large-scale forest and steppe fires that burned over 26.3 million hectares of forest and steppe vegetation, including pasture lands, causing significant losses of life, property and infrastructure. The loss of forest land in Mongolia is increasing, with severe economic consequences, and a growing realisation that a precious ecological resource which contains virtually all rivers, protects soils and rangelands, and provides essential wildlife habitat.

The Fire Continent Australia: New Vulnerabilities
Australia's fire problems are currently in the focus of fire managers and policy makers in the Australasian region. The continent is facing a dilemma: On the one hand Australia's wildlands have evolved with fire and thus are extremely well adapted to fire. In the late 1990s more than 345,000 wildfires burned an average of ca. 50 million hectares of different vegetation types every year. In many of Australia's wildlands frequent fires of moderate intensity and severity are important to maintain properties and functioning of ecosystems and to reduce the accumulation of highly flammable fuels. Thus, it is generally accepted that fire protection (fire exclusion) in Australia's wildlands will lead to fuel accumulation and, inevitably, to uncontrollable wildfires of extreme behaviour, intensity and severity. These ecosystems need to be burned by natural sources or by prescribed fire.

On the other hand there is a trend of building homes and infrastructures in the wildlands around the Australian cities. This exurban trend of settlement has created new vulnerabilities and conflicts concerning the use of fire as a management tools. The extended wildfires occurring in New South Wales in early and late 2002 did not harm the vegetation but the values at risk at the wildland-residential interface.

Tropical Rainforests

Fires in tropical evergreen forests, until recently, were considered either impossible or inconsequential. In recent decades, due to population growth and economic necessity, rainforest conversion to non-sustainable rangeland and agricultural systems has proliferated throughout the tropics (Mueller-Dombois and Goldammer 1990, UNEP 2002). The slash-and-burn practices involve cutting rainforests to harvest valuable timber, and burning the remaining biomass repeatedly to permanently convert landscapes into grasslands that flourish for a while due to ash fertilisation, but eventually are abandoned as non-sustainable, or to convert rainforest to valuable plantations. Beyond this intentional deforestation there is a further, more recent problem, as wildfires are growing in frequency and severity across the tropics. Fires now continually erode fragmented rainforest edges and have become an ecological disturbance leading to degradation of vast regions of standing forest, with huge ecological, environmental, and economic consequences. It is clear that, in tropical rainforest environments, selective logging leads to an increased susceptibility of forests to fire, and that the problem is most severe in recently logged forests. Small clearings associated with selective logging permit rapid desiccation of vegetation and soils increase this susceptibility to fire. Droughts triggered by the El Niño-Southern Oscillation (ENSO) have exacerbated this problem, and were largely responsible for significant wildfire disasters in tropical rainforests during the 1980s and 1990s.

Tropical rainforests cover ~45% of Latin America and the Caribbean. Between 1980 and 1990, when the first reliable estimates were made, the region lost close to 61 million hectares of forest, 6% of the total forested area. During the 1990-1995 period a further 30 million hectares were lost. The highest rates of deforestation occurred in Central America (2.1% per year) but Bolivia, Ecuador, Paraguay and Venezuela also had deforestation rates above 1% per year, while Brazil lost 15 million hectares of forest between 1988 and 1997.

Landscape fragmentation and land cover change associated with this massive deforestation combine to expose more forest to the risk of wildfire, and fires are increasing in severity and frequency, resulting in widespread forest degradation. This change in tropical fire regimes will likely result in the replacement of rainforests with less diverse and more fire-tolerant vegetation types. Although quantitative area burned estimates are sporadic at best, it is estimated that the 1997-1998 El Niño-driven wildfires burned more than 20 million hectares in Latin America and Southeast Asia. The widespread tropical rainforest wildfires of 1998 have changed the landscape of Latin America's tropical evergreen forests by damaging vast forested areas adjacent to fire-maintained ecosystems, such that fires will likely become more severe in the near future, a fact not yet appreciated by resident populations, fire managers, and policy makers. The Latin American and Indonesian problems have much in common, and indicate the problems in tropical rainforests worldwide.

Smoke pollution from tropical rainforest fires, as is the case in many other fire regions of the world, greatly affects the health of humans regionally, with countless short- and long-term respiratory and cardiovascular problems resulting from the lingering smoke and smog episodes associated with massive wildfires.

The environmental impacts from tropical forest fires range from local to global. Local/regional impacts include soil degradation, with increased risks of flooding and drought, along with a reduced abundance of wildlife and plants, and an increased risk of recurrent fires. Global impacts include the release of large amounts of greenhouse gases, a net loss of carbon to the atmosphere, and meteorological effects including reduced precipitation and increased lightning. A loss of biodiversity and extinction of species is also a major concern.

The economic costs of tropical forest fires are unknown, largely due to a lack of data, but also attributable to the complications of cause and effect: negative political implications definitely discourage full disclosure. These can include medical costs, transportation disruption, and timber and erosion losses. They can also, in the post-Kyoto era, include lost carbon costs which, in the case of the 1998 fires in Latin America, can be crudely estimated at $10-15 billion.

The driving force behind the devastating Indonesian fires of 1982-1983 and 1997-1998 was droughts associated with ENSO, in combination with the exposure of rainforest areas to drought as a result of selective logging. It has been estimated that the overall land area affected by the 1982-1983 fires, in Borneo alone, was in excess of 5 million hectares. In non-ENSO years fires cover 15,000-25,000 hectares. The 1997-1998 fire episode exceeded the size and impact of the 1982-1983 fires. During 1997 large fires occurred in Sumatra, West and Central Kalimantan, and Irian Jaya / Papua New Guinea. In 1998 the greatest fire activity occurred in East Kalimantan. In total, the 1997-1998 fires covered an estimated area in excess of 9.5 million hectares, with 6.5 million hectares burning in Kalimantan alone. These widespread fires, all caused by humans involved in land speculation and large-scale forest conversion, caused dense haze across Southeast Asia for an extended period. Severe respiratory health problems resulted, along with widespread transport disruption, and overall costs were estimated at US$9.3 billion. Carbon losses were particularly severe due to high levels of fuel consumption, particularly in peatlands.

Southern African Savannas and Grasslands

Fire is a widespread seasonal phenomena in Africa (van Wilgen et al. 1997). South of the equator, approximately 168 million hectares burn annually, nearly 17% of a total land base of 1014 million hectares, accounting for 37% of the dry matter burned globally. Savannah burning accounts for 50% of this total, with the remainder caused by the burning of fuel wood, agricultural residues, and slash from land clearing. Fires are started both by lightning and humans, but the relative share of fires caused by human intervention is rapidly increasing. Pastoralists use fire to stimulate grass growth for livestock, while subsistence agriculturalists use fire to remove unwanted biomass while clearing agricultural lands, and to eliminate unused agricultural residues after harvest. In addition, fires fuelled by wood, charcoal or agricultural residues are the main source of domestic energy for cooking and heating.

In most African ecosystems fire is a natural disturbance factor to maintain a dynamic equilibrium of vegetation structure and composition, and to nutrient recycling and distribution. Nevertheless, substantial unwarranted and uncontrolled burning does occur across Africa, and effective actions to limit this are necessary to protect life, property, and fire-sensitive natural resources, and to reduce the current burden of emissions on the atmosphere with subsequent adverse effects on the global climate system and human health. Major problems arise at the interface between fire savannas, residential areas, agricultural systems, and those forests which are not adapted to fire. Although estimates of the total economic damage of African fires are not available, ecologically and economically important resources are being increasingly destroyed by fires crossing borders from a fire-adapted to a fire-sensitive environment. Fire is also contributing to widespread deforestation in many southern African countries.

Most southern African countries have regulations governing the use and control of fire, although these are seldom enforced because of difficulties in punishing those responsible. Some forestry and wildlife management agencies within the region have the basic infrastructure to detect, prevent and suppress fires, but this capability is rapidly breaking down and becoming obsolete. Traditional controls on burning in customary lands are now largely ineffective. Fire control is also greatly complicated by the fact that fires in Africa occur as hundreds of thousands of widely dispersed small events. With continuing population growth and a lack of economic development and alternative employment opportunities to subsistence agriculture, human pressure on the land is increasing, and widespread land transformation is occurring. Outside densely settled farming areas, the clearance of woodlands for timber, fuel wood and charcoal production is resulting in increased grass production, which in turn encourages intense dry season fires that suppress tree regeneration and increase tree mortality. In short, the trend is toward more fires.

Budgetary constraints on governments have basically eliminated their capacity to regulate from the centre, so there is a trend towards decentralisation. However, the shortage of resources forcing decentralisation means there is little capacity for governments to support local resource management initiatives. The result is little or no effective management and this problem is compounded by excessive sectoralism in many governments, leading to uncoordinated policy development, conflicting policies, and a duplication of effort and resources. As a result of these failures, community-based natural resource management is now being increasingly widely implemented in Africa, with the recognition that local management is the appropriate scale at which to address the widespread fire problems in Africa. The major challenge is to create an enabling rather than a regulatory framework for effective fire management in Africa, but this is not currently in place. Community-based natural resource management programs, with provisions for fire management through proper infrastructure development, must be encouraged. More effective planning could also be achieved through the use of currently available remotely sensed satellite products.

These needs must also be considered within the context of a myriad of problems facing governments and communities in Africa, including exploding populations and health (e.g. the AIDS epidemic). While unwarranted and uncontrolled burning may greatly affect at the local scale, it may not yet be sufficiently important to warrant the concern of policy makers, and that perception must be challenged as a first step towards more deliberate, controlled and responsible use of fire in Africa.

Fire-Atmosphere Interactions

Research efforts under the Biomass Burning Experiment (BIBEX) of the International Geosphere-Biosphere Programme (IGBP), International Global Atmospheric Chemistry (IGAC) Project, and a large number of other projects in the 1990s were successful in sampling and determination of fire emissions and the identification of emission factors. However, global and regional emission estimates are still problematic, mostly because of uncertainties regarding amounts burned.

Most recent estimates indicate that the amount of vegetative biomass burned annually is in the magnitude of 9200 Teragram (dry weight), i.e., 9.2 billion tons (Andreae and Merlet 2001, Andreae 2002).

These vegetation fires produce a range of emissions that influence the composition and functioning of the atmosphere. The fate of carbon contained in fire-emitted carbon dioxide (CO_2) and other radiatively active trace gases is climatically relevant only when there is no regrowth of vegetation – e.g. deforestation or degradation of sites. NO_x, CO, CH_4, and other hydrocarbons are ingredients of smog chemistry, contribute to tropospheric ozone formation and act as "greenhouse gases". Halogenated hydrocarbons (e.g. CH_3Br) on the other side have considerable impact on stratospheric ozone chemistry and contribute to ozone depletion.

Fire-emitted aerosols influence climate directly and indirectly. Direct effects include (a) backscattering of sunlight into space, resulting in increased albedo and a cooling effect, and (b) absorption of sunlight which leads to cooling of the Earth's surface and atmospheric warming. As a consequence convection and cloudiness are reduced as well as evaporation from ocean and downwind rainfall. The key parameter in these effects is the black carbon content of the aerosol and its mixing state.

Indirect effects of pyrogenic aerosols are associated with cloud formation. Fire-emitted aerosols lead to an increase of Cloud Condensation Nuclei (CCN) that are functioning as seed for droplet formation. Given a limited amount of cloud water content this results in an increase of the number of small droplets. As a first consequence clouds become whiter, reflect more sunlight, thus leading to a cooling effect. As a second indirect effect of "overseeding" the overabundance of CCN coupled with limited amount of cloud water will reduce the formation of droplets that are big enough (radius ~14 μm) to produce rain; consequently rainfall is suppressed. In conclusion it can be stated that

- The fire and atmospheric science community has made considerable progress at determining emission factors from vegetation fires
- Global and regional emission estimates are still problematic, mostly because of uncertainties regarding amounts of area and vegetative matter burned
- Fire is a significant driver of climate change (as well as a human health risk)
- Fire, climate, and human actions are highly interactive
- Some of these interactions may be very costly both economically and ecologically

Forest Fires, Climate Change and Carbon: The Boreal Forest Example

Growing exploitation of the global boreal zone cannot be accomplished without a reconciliation, and compromise, with the fact that the boreal forest is dependent on periodic natural disturbance (fire, insects, disease) in order to exist (Stocks 2001). Forest fire is the dominant disturbance regime in boreal forests, and is the primary process which organises the physical and biological attributes of the boreal biome over most of its range, shaping landscape diversity and influencing energy flows and biogeochemical cycles, particularly the global carbon cycle since the last Ice Age. Human settlement and exploitation of the resource-rich boreal zone has been accomplished in conjunction with the development of highly efficient forest fire management systems designed to detect and suppress unwanted fires quickly and efficiently. Over the past century people throughout northern forest ecosystems have, at times somewhat uneasily, coexisted with this important natural force, as fire management agencies attempted to balance public safety concerns and the industrial and recreational use of these forests, with costs and the need for natural forest cycling through forest fires. Canadian, Russian, and American fire managers have always designated parts of the boreal zone, usually in northern regions, as "lower priority" zones that receive little or no fire protection, since fires occurring there generally have little or no significant detrimental impact on public safety and forest values.

While humans have had some influence on the extent and impact of boreal fires, fire still dominates as a disturbance regime in the boreal biome, with an estimated 5-20 million hectares burning annually in this region. Canada and Alaska, despite progressive fire management programs, still regularly experience significant, resource-stretching fire problems, with 2-8 million hectares, on average, burning annually. In contrast, Scandinavian countries do not seem to have major large fire problems, probably due to the easy access resulting from intensive forest management over virtually all of the forested area of these countries. Russian fire statistics are available over the past four decades but, until recent years, these statistics are considered very unreliable. However, based on recent remote sensing data, it appears that the annual area burned in Russia can vary between 2 and 15 million hectares/year.

Boreal forest fires are, most often, crown fires-high-intensity events that combine high spread rates with significant levels of fuel consumption to generate significant fire intensity and energy release rates. When sustained over an extended afternoon burning period each day, this results in the development of towering convection columns reaching the upper troposphere/lower stratosphere, with significant long-range transport potential.

Recent Intergovernmental Panel on Climate Change (IPCC) reports have emphasised the fact that climate change is a current reality, and that significant impacts can be expected, particularly at northern latitudes, for many decades ahead. Model projections of future climate, at both broad and regional scales, are consistent in this regard. An increase in boreal forest fire numbers and severity, as a result of a warming climate with increased convective activity, is expected to be an early and significant consequence of climate change. Increased lightning and lightning fire occurrence is expected under a warming climate. Fire seasons are expected to be longer, with an increase in the severity and extent of the extreme fire danger conditions that drive major forest fire events. Increased forest fire activity and severity will result in shorter fire return intervals, a shift in forest age class distribution towards younger stands, and a resultant decrease in terrestrial carbon storage in the boreal zone. Increased fire activity will also likely produce a positive feedback to climate change, and will drive vegetation shifting at northern latitudes. The boreal zone is estimated to contain 35-40% of global terrestrial carbon, and any increase in the frequency and severity of boreal fires will release carbon to the atmosphere at a faster rate than it can be re-sequestered. This would have global implications, and must be considered in post-Kyoto climate change negotiations.

Increased protection of boreal forests from fire is not a valid option at this time. Fire management agencies are currently operating a maximum efficiency, controlling unwanted fires quite effectively. There is a law of diminishing effects at work here though, as increasing efficiency would require huge increases in infrastructure and resources. While it is physically and economically impossible to further reduce the area burned by boreal fires, it is also not ecologically desirable, as fire plays a major and vital role in boreal ecosystem structure and maintenance. Given these facts, it would appear that, if the climate changes as expected over the next century, northern forest managers will have to constantly adapt to increasing fire activity. The likely result would be a change in protection policies to protect more valuable resources, while permitting more natural fire at a landscape scale.

Fire Emissions and Human Health

Smoke from burning of vegetative contains a large and diverse number of chemicals, many of which have been associated with adverse health impacts (Goh et al. 1999). Nearly 200 distinct organic compounds have been identified in wood smoke aerosol, including volatile organic compounds and polycyclic aromatic hydrocarbons. Available data indicate high concentrations of inhalable particulate matter in the smoke of vegetation fires. Since particulate matter produced by incomplete combustion of biomass are mainly less than 1 micrometers in aerodynamic diameter, both PM_{10} and $PM_{2.5}$ (particles smaller than 2.5 micrometers in aerodynamic diameter) concentrations increase during air pollution episodes caused by vegetation fires. Carbon monoxide and free radicals may well play a decisive role in health effects of people who live and/or work close to the fires.

Inhalable and thoracic suspended particles move further down into the lower respiratory airways and can remain there for a longer period and deposit. The potential for health impacts in an exposed population depends on individual factors such as age and the pre-existence of respiratory and cardiovascular diseases and infections, and on particle size. Gaseous compounds ad- or absorbed by particles can play a role in long-term health effects (cancer) but short-term health effects are essentially determined through particle size. Quantitative assessment of health impacts of air pollution associated with vegetation fires in developing countries is often limited by the availability of baseline morbidity and mortality information. Air pollutant data are of relatively higher availability and quality but sometimes even these data are not available or reliable.

Vegetation fire smoke sometimes even overlies urban air pollution, and exposure levels are intermediate between ambient air pollution and indoor air pollution from domestic cooking and heating. Because the effects of fire events are nation- and region-wide, a "natural" disaster can evolve into a more complex emergency, both through population movement and through its effects on the economy and security of the affected countries. The fire and smog episodes in South East Asia during the El Niño of 1997-98, in the Far East of the Russian Federation in 1998, and again in Moscow Region in 2002 are striking examples.

The World Health Organization (WHO), in collaboration with the United Nations Environmental Programme (UNEP) and the World Meteorological Organization (WMO), has issued comprehensive guidelines for governments and responsible authorities on actions to be taken when their population is exposed to smoke from fires (Schwela et al. 1999). The Guidelines give insights into acute and chronic health effects of air pollution due to biomass burning, advice on effective public communications and mitigation measures, and guidance for assessing the health impacts of vegetation fires. They also provide measures on how to reduce the burden of mortality and preventable disability suffered particularly by the poor, and on the development and implementation of an early air pollution warning system.

Global Observation and Monitoring of Wildland Fires

From the changing role of fire in the different vegetation zones described above it can be concluded that fire management is becoming increasingly important with respect to global issues for resource management, disaster reduction and global change. With this increasing importance comes the need for a concerted effort to put in place the international global observation and monitoring systems needed to give early warning and identify disastrous fire events, inform policy making and to support sustainable resource management and global change research (Justice 2001). The observation systems will need to include both ground based and space based monitoring components. Advances in information technology now make it easier to collect and share data necessary for emergency response and environmental management. Current satellite assets are underutilised for operational monitoring and fire monitoring falls largely in the research domain. Increasing attention needs to be given to data availability, data continuity, data access and how the data are being used to provide useful information.

There is no standard *in-situ* measurement/reporting system and national reporting is extremely variable and inadequate to provide a regional or global assessment. It is also often hard to relate the satellite and *in-situ* data reporting. Reliable information is needed to inform policy and decision making. Management policies ought to be developed based in part on a scientific understanding of their likely impacts. Fora are needed for exchange of information on monitoring methods, use of appropriate technology, policy and management options and solutions and we need a continued and informed evaluation of existing monitoring systems, a clear articulation of monitoring requirements and operational prototyping of improved methods.

The Global Observation of Land Dynamics (GOLD) project is part of the Global Terrestrial Observing System (GTOS) is designed to provide such a forum. GOLD was formerly known as the Global Observation of Forest Cover (GOFC) project but has been expanded to include non-forested area. The GTOS, which is sponsored by the International Global Observing System Partners, has its Secretariat at the FAO in Rome. The GOLD Program is an international coordination mechanism to enhance the use of earth-observation information for policy and natural resource management. It is intended to link data producers to data users, to identify gaps and overlaps in observational programs and recommend solutions. The program will provide validated information products, promote common standards and methods for data generation and product validation and stimulate advances in the management and distribution of large volume datasets. Overall it is intended to advance our ability to obtain and use environmental information on fires and secure the long-term observation and monitoring systems.

The Design Phase for GOLD is now moving into implementation. GOLD has three implementation teams: fire, land cover and biophysical characterisation. The principal role of GOLD is to act as a coordinating mechanism for national and regional activities. To achieve its goals GOLD has developed a number of regional networks of fire data providers, data brokers and data users. These networks of resource managers and scientists provide the key to sustained capability for improving the observing systems and ensuring that the data are being used effectively. GOLD regional networks are being implemented through a series of regional workshops. These regional network workshops are used to engage the user community to address regional concerns and issues, provide a strong voice for regional needs and foster lateral transfer of technology and methods within and between regions. Networks are currently being developed in Central and Southern Africa, Southeast Asia, Russia and the Far East and South America.

The GOLD-Fire program has a number of stated goals:

- To increase user awareness by providing an improved understanding of the utility of satellite fire products for resource management and policy within the United Nations and at regional national and local levels.
- To encourage the development and testing of standard methods for fire danger rating suited to different ecosystems and to enhance current fire early warning systems.
- To establish an operational network of fire validation sites and protocols, providing accuracy assessment for operational products and a test bed for new or enhanced products, leading to standard products of known accuracy.
- To enhance fire product use and access for example by developing operational multi-source fire and GIS data and making these available over the Internet.

- To develop an operational global geostationary fire network providing observations of active fires in near real time.
- To establish operational polar orbiters with fire monitoring capability. Providing a) operational moderate resolution long-term global fire products to meet user requirements and distributed ground stations providing enhanced regional products. These products should include fire danger, fuel moisture content, active fire, burned area and fire emissions. b) operational high resolution data acquisition allowing fire monitoring and post-fire assessments.
- To create emissions product suites, developed and implemented providing annual and near real-time emissions estimates with available input data.
- It is particularly, important to improve the quality, scope, and utility of GOFC-Fire inputs to the various user communities through:
  - gaining a better understanding of the range of users of fire data, their needs for information, how they might use such information if it was available, and with what other data sets such information might be linked;
  - increasing the awareness of users with respect to the potential utility of satellite products for global change research, fire policy, planning and management; and
  - based on ongoing interaction with representatives of the various user communities developing enhanced products.

Conclusions: Policy and Wildland Fire Science

The challenge of developing informed policy that recognises both the beneficial and traditional roles of fire, while reducing the incidence and extent of uncontrolled burning and its adverse impacts, clearly has major technical, social, economic and political elements. In developing countries better forest and land management techniques are required to minimise the risk of uncontrolled fires, and appropriate management strategies for preventing and controlling fires must be implemented if measurable progress is to be achieved. In addition, enhanced early warning systems for assessing fire hazard and estimating risk are necessary, along with the improvements in regional capacity and infrastructure to use satellite data. This must be coupled with technologies and programs that permit rapid detection and response to fires.

A better understanding by both policy-makers and the general population of the ecological, environmental, socio-cultural, land-use and public-health issues surrounding vegetation fires is essential. The potential for greater international and regional co-operation in sharing information and resources to promote more effective fire management also needs to be explored. The recent efforts of many UN programmes and organisations are a positive step in this direction, but much remains to be accomplished.

In the spirit and fulfilment of the 1997 Kyoto Protocol to United Nations Framework Convention on Climate Change, the 2002 World Summit for Sustainable

Development (WSSD) and the UN International Strategy for Disaster Reduction (ISDR), there is an obvious need for more reliable data on fire occurrence and impacts. Remote sensing must and should play a major role in meeting this requirement. In addition to the obvious need for improved spaceborne fire-observation systems and more effective operational systems capable of using information from remote sensing and other spaceborne technologies, the remote sensing community needs to focus its efforts more on the production of useful and meaningful products.

Finally, it must be underscored that the traditional approach in dealing with wildland fires exclusively under the traditional forestry schemes must be replaced in future by an inter-sectoral and interdisciplinary approach. The devastating effects of many wildfires are an expression of demographic growth, land-use and land-use changes, the socio-cultural implications of globalisation, and climate variability. Thus, integrated strategies and programmes must be developed to address the fire problem at its roots, at the same time creating an enabling environment and develop appropriate tools for policy and decision makers to proactively act and respond to fire.

What are the implications of these conclusions on fire science? Back in the early 1990s the first major inter-disciplinary and international research programmes, including inter-continental fire-atmosphere research campaigns such as the Southern Tropical Atlantic Regional Experiment (STARE) with the Southern Africa Fire-Atmosphere Research Initiative (SAFARI) in the early 1990s (JGR 1996), have clearly paved the way to develop visions and models for a comprehensive science of the biosphere. At the beginning of the 3rd millennium it is recognised that progress has been achieved in clarifying the fundamental mechanisms of fire in the global environment, including the reconstruction of the prehistoric and historic role of fire in the genesis of planet Earth and in the co-evolution of the human race and nature.

However, at this stage we have to ask for the utility of the knowledge that has been generated by a dedicated science community. We have to ask this at a time when it becomes obvious that fire seems to play a major role in the degradation of the global environment. Following the above-cited statement of Pyne (2001) *"Fire has the capacity to make or break sustainable environments. Today some places suffer from too much fire, some from too little or the wrong kind, but everywhere fire disasters appear to be increasing in both severity and damages"* – does wildland fire become a major environmental threat at global level? Does wildland fire at global scale contribute to an increase of exposure and vulnerability of ecosystems to secondary/associated degradation and even catastrophes?

The regional analyses provided in this paper reveal that environmental destabilisation by fire is obviously accelerating. This trend goes along with an increasing vulnerability of human populations. Vice-versa, humans are not only affected by fire but are the main causing agent of destructive fires, through both accidental, unwanted wildfires, and the use of fire as a tool for conversion of vegetation and reshaping whole landscapes.

This trend, however, is not inevitable. There are opportunities to do something about global fire because – unlike the majority of the geological and hydro-mete-orological hazards – wildland fires represent a natural hazard which is primarily human-made, can be predicted, controlled and, in many cases, prevented.

Here is the key for the way forward. Wildland fire science has to decide its future direction by answering a number of basic questions: What is the future role of fundamental fire science, which is the added value of additional investments? What can be done to close the gap between the wealth of knowledge, methods and technologies for sustainable fire management and the inability of humans to be in control?

From the perspective of the author the added value of continuing fundamen-tal fire science is marginal. Instead, instruments and agreed procedures need to be identified to bring existing technologies to application. Costs and impacts of fire have to be quantified systematically to illustrate the significance of wildland fire management for sustainable development.

This implies that the scientific focus has to be shifted. The fire domain for long time has been governed by inter-disciplinary natural sciences research. Engineering research has contributed to a high level of development in the industrial countries. What is needed in future is a research focus at the interface between the human dimension of fire and the changing global environment. The new fire science in the third millennium must be application-oriented and understood by policy makers.

This recommendation is reflected by the establishment of a dedicated Working Group on Wildland Fire of the United Nations. The Working Group operates in support of the Inter-Agency Task Force for Disaster Reduction of the United Nations International Strategy for Disaster Reduction (ISDR) and brings together an international consortium of UN agencies and programmes, representatives from natural sciences, humanities, fire management agencies and non-government organisations (ISDR 2001). The terms of reference of this group is, among others, to advise policy makers at national to international levels in the reduction of the negative effects of fire in the environment, in support of sustainable management of the Earth system. The activities include a major global networking activity – the Global Wildland Fire Network – facilitated through the Global Fire Monitoring Center (GFMC 2003) and supported by the science community.

The UN Working Group also intends to develop a proposal for internationally acceptable criteria, with common procedures and guidelines, for the collection of data on fires in a consistent manner, with the intention of compiling accurate esti-mates of wildland fire globally that can be used by various user communities locally, nationally, regionally and globally.

The contribution of global wildland fire science to the way ahead must lead towards the formulation of national and international public policies that will be harmonised with the objectives of international conventions, protocols and other agreements, e.g., the Convention on Biological Diversity (CBD), the Convention to Combat Desertification (UNCCD), United Nations Framework Convention on Climate Change (UNFCCC) and the UN Forum of Forests (UNFF). The wildland fire community must also search for efficient internationally agreed solutions to respond to wildland fire disasters through international cooperative efforts. The formulation of initiatives on global conferences and summits or through a UN General Assembly for efficient and timely interaction of the international community.

References

Andreae MO (2002) Assessment of global emissions from vegetation fires. Input paper for the Brochure for Policy Makers, UN International Strategy for Disaster Reduction (ISDR) Inter-Agency Task Force for Disaster Reduction, Working Group 4 on Wildland Fire

Andreae MO, Merlet P (2001) Global estimate of emissions from wildland fires and other biomass burning. Global Biogeochemical Cycles 15:955-966

Alexandrian D, Esnault F, Calabri G (2000) Forest fires in the Mediterranean area. UNASYLVA 59(197). http://www.fao.org/docrep/x1880e/x1880e07.htm

American Geophysical Union (ed) (1996) The Southern Tropical Atlantic Regional Experiment (STARE): TRACE-A and SAFARI. J Geophys Res 101(D19): 23519-24330

Davidenko EP, Eritsov A (2003) The fire season 2002 in Russia. Report of the Aerial Forest Fire Service Avialesookhrana. International Forest Fire News 28

Food and Agriculture Organization of the United Nations (FAO) (2001) FRA Global Forest Fire Assessment 1990-2000. Forest Resources Assessment Programme, Working Paper 55. FAO, Rome, pp 189-191. http://www.fao. org.0/forestry/fo/fra/docs/ Wp55_eng.pdf

Goh KT et al. (1999) Health guidelines for vegetation fire events. Background Papers. Published on behalf of UNEP, WHO, and WMO. Institute of Environmental Epidemiology, Ministry of the Environment, Singapore. Namic Printers, Singapore

Goldammer JG (1993) Feuer in Waldökosystemen der Tropen und Subtropen. Birkhäuser-Verlag, Basel/Boston

Goldammer JG (2001a) Africa region fire assessment. In: FRA global forest fire assessment 1990-2000. Forest Resources Assessment Programme. FAO, Rome, Working Paper 55:30-37

Goldammer JG (2001b) Towards international cooperation in managing forest fire disasters in the Mediterranean region. In: Brauch HG et al. (eds) Security and the environment in the mediterranean. conceptualising security and environmental conflicts. Springer, Berlin/Heidelberg/New York

Goldammer JG (2003) The wildland fire season 2002 in the Russian Federation. An assessment by the Global Fire Monitoring Center (GFMC). International Forest Fire News 28

Goldammer JG, Crutzen PJ (1993) Fire in the environment: scientific rationale and summary results of the Dahlem Workshop. In: Crutzen PJ, Goldammer JG (eds) Fire in the environment: the ecological, atmospheric, and climatic importance of vegetation fires. Dahlem Workshop Reports. Environmental Sciences Research Report 13. John Wiley & Sons, Chichester, pp 1-14

Goldammer JG Furyaev VV (eds) (1996) Fire in ecosystems of boreal Eurasia. Kluwer Academic Publishers, Dordrecht

Goldammer JG, Stocks BJ (2000) Eurasian perspective of fire: dimension, management, policies, and scientific requirements. In: Kasischke ES, Stocks BJ (eds) Fire, climate change, and carbon cycling in the boreal forest. Ecological Studies 138. Springer-Verlag, Berlin/Heidelberg/New York, pp 49-65

Global Fire Monitoring Center (GFMC) (2002). http://www.fire.uni-freiburg.de

International Forest Fire News (IFFN). http://www.fire.uni-freiburg.de/gfmc/iffn/iffn.htm

ISDR Working Group on Wildland Fire (2002) UN international strategy for disaster reduction (ISDR) Inter-Agency Task Force for Disaster Reduction, Working Group 4 on Wildland Fire, Report of the Second Meeting, December 3-4 2001, Geneva. http://www.unisdr.org/unisdr/WGroup4.htm

Joint Research Center of the European Commission (JRC) (2002) Global burnt area 2000 (GBA2000) dataset. http://www.gvm.jrc.it/fire/gba2000/index.htm

Justice, CO (2001) Global observation land dynamics (GOLD-Fire): an international coordination mechanism for global observation and monitoring of fires. In: ISDR Working Group on Wildland Fire 2002. UN International Strategy for Disaster Reduction (ISDR) Inter-Agency Task Force for Disaster Reduction, Working Group 4 on Wildland Fire, Report of the Second Meeting, December 3-4 2001, Geneva. http://www.unisdr.org/unisdr/WGroup4.htm

Kajii YS et al. (2003) Boreal forest fires in Siberia in 1998: estimation of area burned and emission of pollutants by advanced very high resolution radiometer data. J Geophys Res 107 (in press)

Kasischke ES, Stocks BJ (eds) (2000) Fire, climate change, and carbon cycling in the boreal forest. Ecological Studies 138. Springer-Verlag, Berlin/Heidelberg/New York

Mueller-Dombois D, Goldammer JG (1990) Fire in tropical ecosystem and global environmental change. In: Goldammer JG (ed) Fire in the tropical biota. Ecosystem processes and global challenges. Ecological Studies 84, Springer-Verlag, Berlin/Heidelberg/New York, pp 1-10

Pyne S (2001) Challenges for policy makers. In: ISDR Working Group on Wildland Fire 2002. UN International Strategy for Disaster Reduction (ISDR) Inter-Agency Task Force for Disaster Reduction, Working Group 4 on Wildland Fire, Report of the Second Meeting, December 3-4 2001, Geneva. http://www.unisdr.org/unisdr/WGroup4.htm

Steffen WL, Shvidenko AZ (eds) (1996) The IGBP Northern Eurasia study: prospectus for integrated global change research. The International Geosphere-Biosphere Program: a study of global change. International Council of Scientific Unions (ICSU), IGBP Stockholm

Shvidenko A, Goldammer JG (2001) Fire situation in Russia. International Forest Fire News 24:41-59

Sukhinin AI et al. (2003) The 2002 fire season in the Asian part of the Russian Federation. International Forest Fire News 28

Stocks BJ (2001) Forest fires, climate change and carbon storage in boreal forests. In: ISDR Working Group on Wildland Fire 2002. UN International Strategy for Disaster Reduction (ISDR) Inter-Agency Task Force for Disaster Reduction, Working Group 4 on Wildland Fire, Report of the Second Meeting, December 3-4 2001, Geneva. http://www.unisdr.org/unisdr/WGroup4.htm

Schwela DH et al (1999) Health guidelines for vegetation fire events. Guideline document. Published on behalf of UNEP, WHO, and WMO. Institute of Environmental Epidemiology, Ministry of the Environment, Singapore. Double Six Press, Singapore. http://www.who.int/peh/air/vegetation_fires.htm

United Nations Environment Programme (UNEP) (2002) Spreading like wildfire. Tropical forest fires in Latin America and the Caribbean. Prevention, assessment and early warning. UNEP Regional Office for Latin America and the Caribbean. http://www.fire.uni-freiburg.de/GlobalNetworks/SouthAmerica/UNEP%20Report%20Latin%20America.pdf

van Wilgen B et al (eds) (1997) Fire in Southern African savannas. Ecological and atmospheric perspectives. The University of Witwatersrand Press, Johannesburg

2.4   Disaster Prevention

Erich J. Plate

Emeritus Professor of Civil Engineering, University of Karlsruhe(TH),
c/o Institut für Wasserwirtschaft u. Kulturtechnik, Kaiserstr. 12, 76128 Karlsruhe

Introduction

The International Decade for Natural Disaster Reduction (IDNDR) from 1990 to 1999 has increased international awareness of the fact that protection from disasters is a basic human right, and lack of preventive measures in many countries in the developing world is a severe impediment to sustainable development. Linkage between disaster prevention and development was clearly stated in the Yokohama Declaration, proclaimed at the end of the Midterm Conference for IDNDR in Yokohama in 1994, and reiterated in the closing forum in Geneva, in 1999, of the IDNDR. Activities of the Decade were based on recognition that technology and indigenous knowledge of coping strategies for natural disasters are available for improving the present state of vulnerability, and that it was more important to apply existing knowledge than to develop new scientific methods or tools. Nevertheless, the decade also challenged scientists and engineers to cooperate in defining and applying appropriate technologies and methods, and to close knowledge gaps by target oriented research and development. It is purpose of this chapter to briefly discuss means by which disasters can be prevented. Although the considerations given here apply in principle to individuals, households, communities or countries, the emphasis is on the political process of disaster mitigation.

Mitigation concept: "Living with risks"

It is appropriate to start with a definition of disaster, and in context of the present book we are talking about natural disasters, i.e. disasters caused by impacts of extreme natural events on a vulnerable society. Society may consist of a population, a person, or a group of persons, for example a community. It is vulnerable because the event may cause harm. If the exposed persons are able to cope with impacts of extreme events we do not speak of a disaster. However, if outside help is needed for overcoming effects of the event, then a disaster occurs. *A disaster is a situation, caused by the impact of an extreme event on a vulnerable population, population group, or person, that exceeds the capability of the affected to cope without external help.* This is one possible definition for a disaster. Other definitions look for the magnitude of losses: a disaster is a situation where damage is higher than a certain arbitrarily defined number of US $, or an arbitrary number of casualties. However, the definition in italics is preferable, because it links disaster with the state of society. This linkage implies that disaster prevention is not only a technical or administrative task, but that it has a strong social component. Disasters are not only events caused by lack of protection, but also by a lack of resistance of the endangered population. Whereas natural events cannot be avoided - i.e. an earthquake in an active tectonic region, or a flash flood in a mountain valley - they do not need to lead to disasters, if the stricken population can cope with its impacts.

In consequence of this definition, it is maintained that *whereas we cannot prevent extreme natural events to cause extensive damage, it is within our means to prevent disasters.* This is an important statement, as it is in opposition to the philosophy of disaster management of the past. People in developed countries were accustomed to see as duty of their governments to protect them against all conceivable threats to life and property. A consequence is that they were not prepared for the extreme event against which there was no protection - every engineering solution has an upper limit of what can be done. As experiences of recent years with extremely disastrous natural events ranging from large floods on Rhine and Mississippi in 1993 to the Kobe earthquake of 1995 and large floods in the Czech Republic, Poland and Germany in 1998 and 2002 have shown, you cannot be prepared against all natural events of any conceivable magnitude. The philosophy of absolute protection has given way to a more modern - and more modest - philosophy of protecting up to a reasonable level by technical or non-technical measures, and to be prepared to handle consequences if an event strikes whose magnitude exceeds the protection level, and to improve the coping capacity of people at risk against the residual risk. This is the concept of "living with risk". The Secretariat of the International Strategy for Disaster Reduction (ISDR), successor to the UN Secretariat for the IDNDR, has published a report with the title "Living with Risk" (ISDR 2002), in which this concept is discussed in detail.

Strengthening coping capacity of endangered populations has many facets. Its purpose is to decrease vulnerability of people, so that they are better capable to recover from impacts of extreme events by their own resources. But what is vulnerability? A clear and universally accepted definition of vulnerability does not yet exist. Let us distinguish social vulnerability: vulnerability associated with social conditions of a person – including loss of life, environmental vulnerability, as a result of the location and exposure of the land, and economic vulnerability, which refers to economic losses that may be caused by the event. Social vulnerability is due to fragility of the personal condition, i.e. higher vulnerability of children and elderly persons in comparison to adults, as well as differences in entitlement (higher vulnerability of the poor than of the rich, of women than of men, etc, Blaikie et al.

1994). Environmental vulnerability has been discussed in previous chapters, it is strongly enhanced by adverse environmental conditions, such as exposure of the land on which people are living (or are forced to live), and by influences of degraded land on the nature of extreme events – a deforested catchment has a higher storm runoff than a forest, and denuded mountain slopes are more prone to landslides than forested ones. Consequently, improvement of environment and poverty alleviation are important and complementary aspects of vulnerability reduction: people often settle in endangered areas because they are too poor to afford living in less endangered regions, and people may degrade the environment as a result of economic conditions, thereby enhancing their own susceptibility to disasters. However, not only the poor face adverse exposures: many well to do people in developed countries select dangerous areas for their dwellings because of threats associated with living in exposed areas are compensated, in their view, by benefits which they prefer: for example, a beautiful view from houses on mountain sides or on sea shores. The third type of vulnerability is economic, it describes losses caused by the impact of an extreme event and is measured in monetary units.

From the point of view of disaster prevention, reduction of total vulnerability by changing social vulnerability is a long term approach. Reduction of environmental vulnerability is a viable alternative, and many non-structural and structural solutions to risk problems are based on changing the (local) environment. A short term approach for increasing coping capacity is by improving preparedness of individual persons through information and training. If already in schools children are trained to react to signs of danger, such as knowing what to do if an earthquake strikes, or if people know where to go if a storm warning is sounded, and in particular if people learn to think clearly and avoid panic, much of the threat from a natural extreme event can be reduced.

The risk management cycle

A systematic approach to disaster prevention is through risk management. Although risk management has many aspects that can be formalised and expressed through mathematical models, it is first of all a systematic way of handling the threat of extreme events: by protection through technical solutions, or by management with the purpose of reducing losses for extreme events that exceed the capacity of the protection system. During recent years, a good understanding of the nature of risk management has evolved, in which all activities related to protection against disasters are encompassed (ISDR, 2002, Plate, 2002). As a result, risk management is a process that consists of a cycle of actions, as schematically shown in Figure 2.4.1.

We notice in Fig. 2.4.1 that the risk management cycle comprises two very different groups of actions: planning and operation. This division reflects the fact that two different groups of experts are involved. The first group are planners and engineers, who are designing, evaluating and implementing a protection system – city

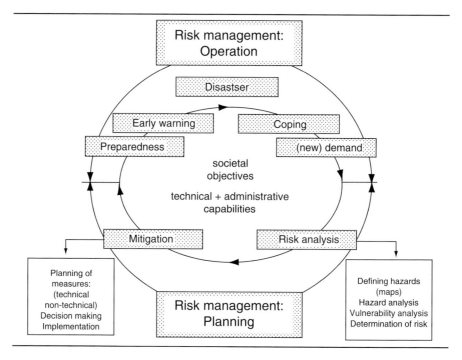

Fig. 2.4.1 The risk management cycle

and regional planners, engineers, and decision makers. The second group are operators, who manage the system – administrators of the protection system, assisted by external supporting groups, such as fire brigades, red cross helpers, etc. Tasks involved are also quite different. For example, a forecasting system is developed by engineers. Once found to be functioning as required, the system is handed over to operators, who use it in their day to day operation.

Discussion of all aspects, both for planning and for operation, of the risk management cycle is the approach taken here to explain different components and actions that are involved in disaster prevention.

The planning phase of risk management

Risk perception

Obviously, the first step in starting the risk management process is through awareness, by authorities and the people, that danger exists of being harmed by an extreme event. This seems to be self evident. However, many examples can be given to show that this is not so. For example, recent research by the Geoforschungszentrum

Potsdam (Bormann, oral communication) has shown that the city of Cologne may be threatened by earthquakes, with a fairly high probability, of which the city administration had not been aware – and very likely will not want to be aware of, as it implies grave consequences if numerous buildings in the city that are not earthquake resistant were to be checked. Many other, actual dramatic disasters can be cited as evidence of a lack of risk awareness, or at least of underestimating dangers from natural sources: unexpected volcano eruptions, such as the 1985 Nevada del Ruiz event, where approximately 23000 people were killed by lahars (mudflows), flash floods in valleys that never experienced flash floods before, earthquakes in areas where earthquakes were not expected even by seismologists, landslides due to permafrost melting of mountain slopes that have not moved in the past (Azzoni et al.1992). More often, awareness is created by occurrence of a disaster. People in Grimma, a city located in Saxony on a tributary of the Elbe, which has been devastated during the large flood in August 2002 (see appendix), had not been aware that the small river flowing through the city could turn into a torrent. However, communities usually enter the risk management cycle knowing that a large extreme event has occurred or could occur any time, against which protection is being sought.

Already during reconstruction activity after a disaster, better protection against the kind of disaster that has caused the destruction should be introduced, preferably by reducing vulnerability – but unfortunately it seldom is. Disaster help most of the time is used to reestablish the "status quo ante", the situation as it was before – often with the feeling that the large extreme event had been a once in a 100 years event, so that the area would have a respite for another 100 years. As an example, after the big flood on the Oder river in 1997, (Grünewald et al., 1998) the river which forms the border between Poland and Germany), it was determined that it would be much cheaper to relocate houses of people living in the flooded part of the Oderbruch (on the West side of the river) – and that money was available to do that – than to rebuild the houses. But people preferred to rebuild, so that similar damage may be expected in the future, if dikes are not raised at much higher cost.

Planning for protection: risk assessment

The decision to improve protection starts the planning phase of the risk management cycle. To prepare for planning, data have to be collected and reviewed, in order to identify sources of threats and evaluating their disaster potential, i.e. magnitudes and effects of threatening events. The rational approach to determine the disaster potential is risk assessment. Risk assessment is the activity by which threats of all extreme events that can be expected to occur in a certain region are assessed and their consequences evaluated. In simplest terms risk assessment is done by looking back into the past: the concept often followed is to be prepared to handle the largest event that had occurred in historical past, using as design concept that what has happened yesterday either in our own or in neighboring regions should never again happen. For example, flood protection works are frequently designed to protect against the largest flood on record. Although this approach is not very scientific, it is better than just to rebuild what has been destroyed and hope that the region will not be affected again.

Risk assessment as a modern tool proceeds in a more rigorous fashion (WMO 1999). The first activity is to obtain estimates of extreme events of the future. As the extreme event is (usually) a rare event, the degree of a threat from an extreme event is not described only by its possible magnitude, but also on a measure of frequency of occurrence – an extreme event that occurs only once in every 1000 year period, on average, is not as threatening as a less extreme event that occurs every few years. Therefore, risk assessment uses the distribution of extreme events in combination with the probability that they will be reached or exceeded. In engineering literature, this combination is called the hazard for the event.

Hazard determination from data

Hazards from extreme natural events and their impacts are analysed in detail for larger, engineered projects, or where, as in the cases cited for volcanoes and avalanches, lives are directly endangered. The approach for determining hazards is based on statistical evaluation of past or potential occurrences. The best statistical basis consists of time series of events. This information is provided by natural scientists: seismologists determine statistics of earthquake magnitudes and of tsunami occurrences, meteorologists for extreme winds, icing events and extreme rain- and snow falls, hydrologists for extreme floods, oceanographers for storm surges. Time series are used to identify annual extrema and to determine their probability distribution. The process of determining probability distributions for extreme values has its longest tradition in hydrology. Extreme value analysis, (see for example, Ang and Tang 1984, Plate 1992, Kottegoda and Rosso 1997) was developed in particular for obtaining design floods for given exceedance probabilities.

Because extreme value statistics is faced with many problems, there exists a very large body of literature on the subject. Only three problem areas may be mentioned. First, for most types of extreme values of natural events there exists no physically based mathematical function to describe the shape of the probability function, and yet, in order to obtain estimates for rare extreme events, the uncertain probability curve has to be extrapolated to values that may never have been observed. Statisticians know well that inferring shapes of probability distributions from data can lead to very misleading results.

A second dominant problem of extreme value statistics is that usually time series which form the data base, are fairly short, and are to be extrapolated to extremes for which no, or insufficient, data are available. Bias corrections must be considered, and uncertainty estimates of the so called T-year event – an extreme event that is exceeded once every T years on average. Consequently, design parameters must be inferred from a limited number of sample points, such as inferring the flood that may be exceeded once every hundred years from a thirty year record, or you have to obtain the probability distribution of earthquake magnitudes from a small number of earthquake events observed locally. Therefore, there exists a fairly wide margin of uncertainty for design estimates (Kottegoda and Rosso 1997).

In recent years an added problem has entered: considerations of non-stationarities in time series of extremes, as may be introduced by changing land use, or by climate change. The significance of such a change is illustrated in Figure 2.4.2. Assume that a change does nothing more than increase average $\bar{u}$ of extreme values u by a small amount, without changing shape or other parameters of the probability distribution. It can be seen that this can have a very large effect on exceedance probabilities. In the example shown, by $\bar{u}_1$ shifting of today to $\bar{u}_2$ due to climate change, event u_T which today is the event exceeded once every T= 1000 years, is shifted to become the value corresponding to T = 100. More extreme events may be influenced even stronger, in particular as it can be expected that variability of extreme events is also increased, so that with climate change the tail of the probability distribution of Figure 2.4.2 would move even further to right.

Developing methods of extending time series of extreme events beyond historical records is an important research topic. Some extreme events, such as volcano eruptions or earthquakes of a very distant past have left their traces in geological formations, droughts in sea bottom deposits or in tree rings, or floods left marks of debris in caves high up on river banks in canyon walls, indicating ancient flood levels. Determination of ancient floods is an important task in areas, where no or only very few measured data are available, and where large floods need to be considered, as for designing the spillway of a large dam. Another example is inferring drought statistics in semi-arid regions from tree rings and lake deposits (WMO, 1999), as is done for the state of Arizona, USA.

Improving statistics based on earlier occurrences is only one means of getting information for identifying danger from natural events. Potential landslides can be inferred from a careful analysis of the geology of a region, and from indicators such as small ground dislocations or from trees inclined to vertical (Dikau et al. 1996). Avalanches may happen not only in locations with history of previous occurrences, but also from large snow accumulations elsewhere on steep mountain slopes.

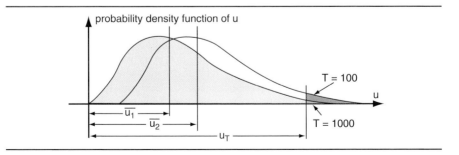

Fig. 2.4.2 Effect of a shift in probability due to an increase of mean value

Volcanologists (Schminke 2001) use the approach of identifying volcanoes by type and determine the potential of an active volcano to erupt from its past history and by comparison with other volcanoes of same type.

Standards for design

In general it is not possible to do a risk analysis for all kinds of communities and buildings. Engineers have learned to design against extreme loads by developing design standards: codes which embody experience of the most experienced engineers of a country. Design standards for wind or earthquake forces are in force in most countries. Standards are an acceptable compromise between maximum conceivable event and economics. Engineers have always realised that absolute protection cannot be obtained, and they therefore set limits in their standards of what is to be demanded in terms of building safety, and what is beyond economic limits. Design loads are not given for maximum conceivable loads, they correspond to extreme events with certain recurrence intervals, for example, for events that may be exceeded only once every 30 years (as is usual for wind loads) or once every 100 or two hundred years, as is not uncommon for rivers.[1]

Design standards for environmental loads rely on maps, in which the degree of threat from an extreme event of a certain type is mapped. Typical are maps of wind zones, in which the "design wind" is identified, which should be used in conjunction with appropriate building type-specific load formulae to determine wind forces on buildings. Recent building code specifications for wind loads include topographic wind adjustment factors for mountain slopes or for city complexes. Other types of maps are maps for identifying danger from earthquakes. They show areas that are particular prone to earthquakes and identify a "design earthquake", to be used in design of engineered structures (Ahorner and Rosenhauer 1993, Grünthal et al.1998).

For flood estimation, the US Water Resources Council used maps to regionalise parameters of the extreme value distribution. By analysing a large number of gauges

---

[1] Design information of protection against, say, the one hundred year event was often misunderstood to mean that once the one hundred year event has occurred, you will have peace for another 100 years. What really is implied by using the term "one hundred year event" is that the probability that it is reached or exceeded in any one year at a particular point in space is 1%. This probability changes if you look at a region. It increases if the region is large. For example, it is rather likely that every year we will have a once in 100 year flash flood event in one or more catchments somewhere in Germany, or there exists a larger probability of earthquake occurrence for California than for San Francisco. This is an indication of events being associated with areas which are statistically independent. What is measured locally is representative of a certain area, whose size is determined by regional correlations. Regional correlations of events, for example for rainfall events, or for extreme storms, or for earthquakes are not well understood for most regions, and development of methods for this purpose is an important research task. For an example of a recent study on this issue see Madsen and Rosbjerg (2002).

on many different rivers in the USA it was found empirically that the log-Pearson III distribution (better known as three parameter logarithmic gamma distribution) – gives a good fit to most observed data, and that skew coefficients of this probability distribution (pdf) could be regionalised (Chow et al.,1988). Mean value, standard deviation, and skew coefficients are sufficient to fully specify the log-Pearson III distribution. Maps of skew parameters have been prepared for the whole USA. In this way, the flood estimation procedure was codified and became legally binding. In similar ways, standards[2] have been developed for other types of hazards, such as extreme winds and earthquakes.

In flood protection it is usually not sufficient to know the magnitude of the flood determined from such regionalisations. For most flood protection measures discharges need to be converted into flow depth - be it for designing the spillway of a dam, or for determining the height of dikes along a river, for which the flood must be converted by hydraulic calculations to obtain the water level for the modified cross section of the diked river. Flood events for rivers, with or without dikes, are caused by overtopping of banks and flooding of low lying areas. Therefore, discharges have to be converted to water levels, which are used in conjunction with topographic maps to produce flood hazard maps, showing flooded areas for different exceedance probabilities (WMO 1999). Such maps are critical documents, as property values may depend on them. And because they depend strongly on local topography, much controversy can develop from inexact topographic maps. Therefore, flood hazard maps, if they are used at all, are converted into so-called "flood risk maps", as for example used in Switzerland, in which flood exposure is expressed in colors, usually in three or four classes: "high risk" corresponding to flooding levels to be expected about every ten years, to "low risk" areas where flooding can be expected to occur less frequently than once every 100 years.

Maps of the kind described form the basis for deciding on what protection approach to use. In the simplest case, nothing is done to improve the situation, but people are made aware of problems, and reduction of vulnerability is the only course of action, if any, that is taken. If protection at a higher level is sought, technical or non technical solutions are options that need be considered.

In the past, an engineered building that has been designed according to rules of the trade, and where design loads include factors of safety that reflect experiences of generations of engineers, was considered as absolutely safe. But experience

---

[2] The principle followed in setting standards is that in designing structures, it is less important that design methods are physically correct and accurate (although engineers and natural scientists who draw up codes usually have a very good understanding of the basis of the codes, and tend to incorporate solid research findings – but codes are not changed very often and usually reflect the state of the art of many years ago), but that they are applicable as upper reasonable limit to a very wide range of conditions. They relieve the engineer from the arduous task of finding his own design loads, and they protect him against litigation. However, usually standards only give recommendations, and for good physical reasons, such as newer research findings, the engineer may decide against using codes.

has shown that extreme events may occur that surpass the design load, and therefore a more recent change in design paradigm is to design also for the case of failure of buildings. Much engineering research today goes into finding ways of making buildings to collapse safely, i.e. in such a way that people can escape from a collapsing building – the principle of designing safe against failing is replaced by the principle of safe failing. For example, dams in large scale water projects today are not considered absolute safe, but you also investigate what happens if the dam should fail (Betamio de Almeida and Viseu 1997).

Vulnerability determination

Conventional engineering is not concerned with vulnerability. According to the simplistic concept of design, a structure that is well designed according to standards cannot fail. To allow for imperfections and other uncertainties, a substantial factor of safety is used. The case that design loads may be exceeded, for example by storms exceeding the design wind speed, is not considered. Unfortunately, the existence of codes does not prevent buildings from collapsing, for example for sub-design earthquakes. Much too often you encounter poor workmanship, inadequate materials, or poor maintenance of structures. During the IDNDR, engineers therefore have pointed to the need of enforcing standards during design, and quality supervision during construction.

A more sophisticated approach to design includes considerations of economic vulnerability of each of the buildings exposed to extreme events. A broader definition of economic vulnerability includes costs resulting from the failure of the structures, where failure is defined as inability of the structure or more general, of the protection system, to meet the demand for which it is designed.

In vulnerability assessment, you have to distinguish operational and structural failure. Operational failure refers to a failure which interrupts operation, i.e. a forecasting system may fail due to a power failure. Failure of structures can occur because design loads are exceeded – for example, a building designed to withstand an earthquake of strength 5.5 on the Richter scale fails if the actual earthquake is of strength 5.9. Failure costs include both costs of replacing or repairing damaged structures, and costs of damage caused by failure.

Social vulnerability needs also to be considered, which expresses the ability of people to manage natural events and failure of existing protection systems, as well as the susceptibility of different groups of people – especially the poor – to be harmed by them. An exact quantification of social vulnerability in this sense is not available. It is strongly affected by environmental vulnerability, caused by degradation of environment or by pressure of increases in population. The World Bank has recently drawn attention to the interrelation of vulnerability and environment and has made vulnerability reduction against environmental hazards part of its environmental strategy (World Bank 2002, see table 2.4.1).

Risk determination

Combining vulnerability and hazard into risk is the last part of the risk assessment process. This task is comparatively easy, if risk is to be quantified in monetary terms, i.e. if costs and benefits are given in monetary units. Costs of construction and benefits from avoided damage are considered, as well as the residual risk, which quantifies cost of damages which occur should the protection system fail – damages both to the protection system, and to the elements at risk.

It is evident that planning a protection system cannot be based only on the disaster that you have just experienced. Rather, you have to use all information to make a probabilistic assessment of all probable extreme events, and to combine them into decision criteria which permit to evaluate benefits and costs of each of possible alternatives for meeting protection objectives. This is reflected in the residual risk, or risk cost, which is the expected value of costs which would be incurred, on average, during lifetime of the structure, or more generally, during lifetime of the threatened object or person – called "element at risk" in insurance terminology. If the damage is the same for all extreme events above a certain threshold, risk cost is the product of hazard (expressed as exceedance probability) and cost of replacing the element at risk – a building or a structure. In order to allow for a certain reduction of damage for lesser events, risk costs in an exact cost benefit analysis are defined as integrals over all costs that could occur if all possible extreme events would be evaluated both with their consequences and their probability of occurrence. Insurers allow for average damage to elements at risk for all extreme events by an average factor, called exposure factor Ex, and write:

$$RI = Ex \cdot K \cdot P$$

where RI is the risk, the exposure factor Ex is a number between 0 and 1, K is the maximum damage that can occur, i.e. the replacement value of the element, and P is the exceedance probability for the extreme event, i.e. the probability that an event occurs that is equal to or larger than a certain critical value. Risk cost is the quantity on which insurers base their calculations of premiums (i.e. Munich Re, 2002).

Risk based on economic vulnerability alone is not sufficient as a decision criterion. A damage of 1 Mio US $ has an entirely different meaning for a poor country or community than for a rich country or community. Therefore, in the political decision process decision criteria include not only monetary risks, but also social and environmental vulnerabilities. For example, the value of saving human lives, or of protecting the environment must also be considered, and these are not measurable in terms of money - all attempts to assign a monetary value to human lives have failed in the past, and it is not likely that they ever will, except in a very arbitrary scale of values. Therefore, risk costs and risk to life and health need to be determined independently. Theoretically, the best approach is to permit only those alternatives which minimise loss of life and do not impair the environment, and to select the optimum solution from the remaining alternatives.

Mitigation

Mitigation consists of all actions within the planning and implementation process that lead from risk assessment to the final protection system. Actions for improving disaster prevention systems can be identified by three possible approaches: mitigation, no action, or preparedness. Mitigation is the activity of preparing a permanent protection system or strategy, whereas preparedness is the set of precautionary non-technical measures taken ahead of time for being prepared against all extreme events that may occur in future. For planning, results of both "slow onset" extreme events, i.e. events that develop gradually in an usually irreversible manner, such as land degradation and deforestation, and of "sudden onset" extreme events, i.e. non-persisting impacts of extreme event, such as earthquakes, floods, droughts, storms and similar causes must be quantified.

There usually are many different options by means of which protection against natural extreme events can be found. In most areas, people have learned to reach a degree of protection by self help. Houses in the Mekong region and in parts of Australia are put on stilts, or houses in North Germany and elsewhere are put on artificial mounds to raise them above flood level. In low lying areas, check valves are installed in pipes connecting houses to sewers to prevent backwater from flooded rivers from backing up into houses. In earthquake prone areas houses have been built with light weight material for roofs, so that a collapsing roof cannot cause excessive damage, and people have learned to fortify corners of buildings to prevent them from collapsing under horizontal forces due to earthquakes or wind, and roofs of buildings have been hipped, to offer less resistance to strong winds. There are a large number of such self help measures, many of them developed locally over many generations - and today planners are required to consider these methods in their design strategy.

Standards are usually only required to be used for engineered construction, not for non-engineered construction. For example, one or two family houses usually are built with little regard to safety against natural extreme events. Limitations are set by the ability of house owners to afford protection. However, more often local traditions are just as important as economics: the situation of hurricane resistance of American houses comes to mind. The light construction usual in the USA is one reason why hurricanes cause such enormous damage. The heavily built stone houses of the French countryside, for example, would very likely suffer much less (Munich Re, 2002).

Mitigation: Technical solutions

In Figure 2.4.3 typical technical systems for flood management are shown, as they are used, for example, in the Rhine catchment. The main river course is parallel to a mountain range, in which tributary rivers originate, which can contribute large discharges to the main river. The classical method of protection is by dike systems. These become particular important if the river has been trained. In the example,

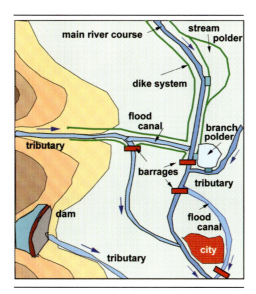

Fig. 2.4.3 Technical solutions to flood disaster mitigation

training is indicated by barrages on the main stem of the river. On the Upper Rhine barrages have been built to increase the water depth for river navigation (and power generation). To prevent tributary floods to flood the river plain, different methods are indicated. One can build dikes along the river, or a reservoir for retaining part of the flood waters is arranged in the upstream part of the catchment of the tributary. Also often used is the method of flood diversion. For example, due to geomorphological processes tributaries of most rivers are running almost parallel to the main stem of the river over considerable distances. These distances can be shortened by introducing flood canals: a barrage is arranged across the course of the tributary, and by means of the flood canal so much water is diverted the shortest way possible to the main river that the remaining river flow does not overflow the banks.

Flood control along the main channel of the main river is accomplished by dikes, as is shown in the upper part of Figure 2.4.3. To prevent flooding of the downstream region, for example of the city complex bottom right of Figure 2.4.3, we have local solutions, consisting of dikes (not shown) and of a diversion canal, by means of which flood waters are bypassed around the city - a solution that is widely used for reduction of floods in large cities which are located near large rivers. Examples are the Danube diversion canal in Vienna, and the bypass for the city of Wroclaw in Poland. Such diversions also require building of weirs or barrages, in order to control the flood in the two branches of the river.

Other protective measures for the city should involve upstream control. Natural and artificial lakes are a good way of controlling some of the flood: by natural means, or, as is the case in alpine lakes, by controlling water level in the lake by means of a weir at the lake outlet. Alternatively, part of the flood volume of water can be diverted into stream polders. In contrast to the polders in the Netherlands or in northern Germany, which are areas surrounded by dikes to keep sea water out, river polders are actually artificial and temporary lakes to keep water in: during flood, water is diverted from the river into polders, and when the flood wave has passed, the water is returned to the main river. To facilitate emptying and filling of the polders, complex hydraulic structures are needed. An elegant solution is to arrange a branch polder at the confluence of two rivers, as for example on the confluence of Rhine and Moder rivers in France near Strasbourg. Here, the Rhine is held back by a barrage (in this case already existing as part of the river training system for navigation) so that its level is high above the surrounding country, and the Moder flows into the Rhine downstream of the barrage. The large difference in water surface elevation of the two rivers facilitates emptying and filling of the polder.

Some additional space can be obtained when gates of barrages are controlled in such a way that some flood water is retained in the section between barrages. However, such an operation can only be successful, if opening and closing of barrage gates is closely attuned to the flood. This requires a very excellent forecast of the flood hydrograph. To provide the forecast information needed for operation of filling and emptying of polders and of gates of weirs for flood control and to convert forecasts into appropriate on-line operation rules is a complex task, and a continuous improvement for forecasting capability is a necessary research target for meteorologists and hydrologists.

Mitigation: Non-technical alternatives

Modern disaster management tends to stay as much as possible away from complicated technical solutions and to rely on less expensive, non-technical – and, as often is implied, more environment-friendly – solutions. The conceptually simplest solution is to move endangered people permanently out of the endangered region. This solution was quite typically sought in antiquity: cities were founded on elevated ground near rivers, to offer the convenience of river travel and water supply without flood hazard. In mountain areas, populations stayed in the lower part, where neither flash floods nor land slides or avalanches could threaten them. Rivers and mountain valleys were left alone. However, land shortage and activities associated with rivers or mountain areas made people to accept the danger in return for advantages gained. Thus, permanent removal of people from endangered areas is usually a non-feasible solution, and, in particular right after a large event, these people usually ask for technical solutions.

On economic grounds, non-technical solutions generally are preferable. There is no question that technical solutions tend to be expensive, and an investment into protective systems that are not needed except once every ten or so years is not attractive, especially in a economic situation where many tasks of improving the infrastructure of a country seem to have higher priority. A technical solution through dikes is absolute necessary, regardless of cost, when the security of large groups of people is endangered by an extreme event: protection of Holland by means of dikes protecting against the 10,000 year extreme storm surge of the North-Sea is a precaution where investment, although huge in monetary terms, is small in comparison to avoided damage. However, when damage is small in comparison with cost of the system, it may well be useful to concentrate on non-technical solutions. Non-technical temporal solutions are designed to save peoples lives, not necessarily their property. The idea is to warn people in real time that an extreme event is forthcoming and provide means for them to escape to safe areas.

The change in attitude – to consider expensive technical solutions only as a last resort – has found a rather remarkable manifestation in the case for flood protection in Bangladesh for people living in the large delta of Ganges, Brahmaputra, and Mengha. These people are frequently threatened by tidal waves from the Bay of Bengal, and from floods in the three large rivers. After disastrous floods in 1987 and 1988 (Jessen 1996) with a high number of fatalities, the World Bank in the 90s coordinated a large scale study for flood protection in this area. The so called "Flood Action Plan" for Bangladesh was drafted, which provided for very large dike systems with numerous gated structures. This system was never implemented. NGOs and Government Agencies cooperated instead in building shelters, where people could go to in case of floods. Thousands of these shelters have been built. Similar solutions are being sought in other parts of the world: after the 1999 flood of the Limpopo in Mozambique the best solution for much of the country was determined not to be construction of dikes, but construction of safe havens – elevated areas, where people could move to and bring their livestock along.

In both these examples, success of protection measures depends strongly on the ability of forecasting extreme events. Non-technical measures have become feasible in many cases because of vast improvement of forecasting capability (Zschau & Küppers 2002), in Bangladesh for Typhoon tracks, in South East USA and the Caribbean for Hurricanes. Forecasting centres for these tasks have been established in the USA and other countries, and Typhoon activity is closely monitored by the Typhoon Committee, located in Manila on the Philippines. In Europe, there are many centres charged with flood forecasting (Homagk 1996), and a well developed and successful forecasting system exists, for example, for river Rhine. It may therefore be stated with some justification that the most important technical development for risk mitigation is development of improved forecasting systems. It is for this reason that the IDNDR set as one of its three targets for the decade, that at the end of the decade "all people should have access to world wide early warning systems".

Development of improved warning systems is still a challenge to the scientific community. It is a fallacy to believe that warning and forecasting are identical. Forecasting is only the first stage in a more general warning process. The best forecast is worthless if it is not accepted and acted upon. Therefore, early warning has to be seen as a process with the elements depicted in Figure 2.4.4. A forecast has to have a good forecasting model, however, it is also necessary to have

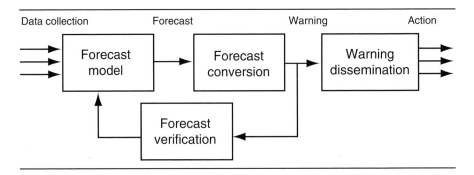

Fig. 2.4.4 Elements of an early warning system.

good data on which to act. For example, for forecasting extreme winds you must have a good meteorological model that can determine in advance how storms will move through the land: not only is the motion of large scale air masses of significance, but also modification of these air masses by topographic features. Even more complicated is the issue of flood forecasting. For large river systems, where flood plains far downstream from headwaters are particularly endangered, a good forecast can be obtained from correlations of downstream gages with upstream gages and rainfall inputs, because the most important effect is downstream translation of flood waves from upstream. For small areas endangered by flash floods the forecast is much more difficult. A detailed rainfall runoff model is needed, which requires good input data from meteorology. Remotely sensed data, taken for example from satellite pictures or from low flying aircraft, or from stationary or satellite mounted radar, have made much better spatial resolution forecasts of rainfall possible, and in combination with ground observation of rainfall they offer very exciting possibilities of improving present day rainfall and storm forecasts. However, for flood forecasting good rainfall forecasts are not enough. In view of the role of soil conditions, they have to be combined with estimates of the state of saturation of the ground. A very wet or even saturated ground will cause much higher runoff than an initially dried out soil.

A good forecast not only allows people to leave unprotected areas, it also can lead to substantial decreases in economic losses. Temporary protection measures can be installed during time between warning and arrival of extreme event. Boarding up of large windows to prevent them from being smashed by hurricane winds is one of the means by which people can protect themselves against hurricane damage, tents can be tied down, other light weight structures fortified. Temporary flood proofing is obtained by flood barriers for entrances and low lying windows. A good method for preventing floors of cellars to buckle under uplift from flood raised groundwater is the technique of filling basements with clean water: this not only prevents basement floors from buckling, but also avoids damage due to polluted water.

Decision processes

Evaluation of financial risks as well as risks to life initiate the decision process, in which alternatives for mitigating potential disasters are studied, and the best, or more accurately, the most acceptable protection system is selected. Obviously, costs of providing a protection system are important, with all its financial and environmental consequences, as well as expected costs of damage when the protective system fails. They have to be compared with monetary, social or environmental benefits that can be gained by being protected up to a desired level. But usually the decision is not only based on economics. The whole value system of society is involved. Therefore, the decision process for finding the optimum solution is a heavily constrained optimisation problem.

Constraints are central to risk management, and therefore they are placed in the centre of the risk management cycle of Figure 2.4.1. One set of constraints is imposed by political objectives: these reflect the value system of a society, - how much protection is felt necessary for society to function, and how much society can and wishes to invest in added protection beyond what already exists. Selection of a protection approach is a political task, involving national priorities, resource allocation and technical and economic capabilities of governments. In developing countries, it may also depend on donor support – and because donor support usually is depending on priorities set by receiving countries, disaster prevention in these countries too is a question of priorities.

Societal constraints include also economical constraints: ability and will of the individual or society to finance a protection project. For example, if protection is sought against large floods, individuals cannot do much to improve their flood protection. Usually, it is a challenge to the solidarity of society. Even if resources are available for funding protection projects, these funds are in competition with other needs of society, such as infrastructure building or social problems.

The second set of constraints is due to technical and administrative capabilities of a society. A wealthy country with a large and well trained work force has many technical possibilities of building and managing complex systems, can afford the cost of maintaining a protective system and a preparedness structure that is needed only once every few years, whereas a poor country with limited educational capacity cannot protect its citizens the same way. Whatever the criteria are by which selection for a protection system is made, they should always include an estimate of the costs involved, including risk cost.

Because of these constraints, encouragement of and guidance for self help - disaster prevention by appropriate preventive actions of affected - are important aspects of all types of risk management.

All these factors must be considered in making the decision of which system to use. The final decision on a protection system is made through the political process. There are two steps involved: the process of integrating the protection concept into general planning for the region under consideration, and the process of involving the people. General planning requires coordination of development of the region's infrastructure through highways, railways and canals. Involving the people is to assure that planning is not done against the wishes of the people. In public hearings on projects, not infrequently issues are brought forth that planners may have overlooked, and planners obtain an opportunity to fully explain their ideas and concepts for the selected alternative. The final decision on any alternative, however, lies with the customer, i.e. the people that pay for the system. If finally the decision leads to a mitigating measure, then the structural solution (such as building protective structures) or non-structural solution (such as setting up an early warning system) chosen is implemented.

Management of risk: Operation

The second part of risk management is the process of operating the protection system. Obviously, there exists a large difference between planning a new and operating an existing protection system. Once the system has been put into operation, other rules hold - and, what is often overlooked in discussions of disaster management, there are entirely different groups of specialists involved in planning or operation. Therefore a distinction must be made between risk management for planning, and risk management for operation – which is risk management in a narrower sense. Risk management for operation is what is done on a day to day basis in disaster prone areas. If you know that a disaster may strike at any time, you should be prepared for it, and this means you have to stock up on medical supplies, food stores etc. and you must be ready to start an early warning procedure.

There are many different activities that can prepare a population for coping with extreme events. During IDNDR, emphasis was given to make sure that schools train children to protect themselves in case of an extreme event, and to react appropriately if a warning has sounded. Warn systems obviously must be maintained and practice is needed for responding appropriately. This of course not only holds for children: in flood or earthquake prone areas everybody must know what to do: emergency routes for fleeing the area, but also for guaranteeing the unhampered movement of emergency vehicles need to be identified and marked.

Conclusion

Disaster prevention is subject of this paper. It has been shown that this does not mean that risks can be avoided altogether: it only implies that you have to learn to live with risk so that the effects of an extreme natural event do not lead to disaster. A balanced approach for disaster prevention is to mitigate the effect of events with large negative consequences through a combination of technical and non-technical approaches – approaches which usually do not lead to an optimisation problem that can be handled by means of classical optimisation theory, such as developed in operations research, but which lead to a political compromise in which opinions and attitudes of people involved in the decision process are as important as cost-benefit analyses. Disaster prevention is a thankless task, as it involves tying up funds in activities that, hopefully, are not often needed. Therefore, disaster mitigation should be considered during every project that may be influenced or may influence natural extreme events. This has been recognised as a challenge not only for every country, but also for projects of development assistance, and the points to be considered in planning any development project should include components of preparing for natural disaster mitigation. This has recently been recognised by the international funding agencies, and a typical set of considerations for development projects is shown in Table 2.4.1, which is part of the development strategy of the World Bank (World Bank 2002).

Because social conditions change, because technical and financial capabilities will change, and because the environment may change it is of extreme importance that disaster prevention is accepted as a continuing task which every generation has to tackle, and for which every generation has to find solutions according to their value systems. This is the reason for making disaster prevention an integral part of sustainable development: to be free of disasters frees creative energies of a nation towards maintaining or improving living conditions for future generations.

A Case Study: The Central European Flood of August 2002

The large floods of August 2002 on Elbe and Danube counted among the largest disasters that ever hit Germany and the Czech Republic. Here is a brief description of what happened, by referring to the map of Figure 2.4.5, which shows Elbe catchment, Elbe river and its main tributaries.

The meteorological situation: we had a very cold and rainy July and early August. The cold air slowly moved to the East, being pushed from the South by a low pressure area originating over the western part of the Mediterranean. The winds from this depression were channelled by the Alps and brought in moist and warm air from the Mediterranean, which upon meeting the cold air over eastern Germany and further east was moved upwards, cooled and dropped its moisture in the form of torrential rains. This meteorological situation corresponded exactly to the one which led to the 1997 disaster on the Odra river – the next larger river further to the east – a disaster that caused enormous damage exceeding 4 billion US $ in Poland, but which spared Germany: only a small region on the northern border to Poland was affected, and losses were less than 300 Million.

This time, the areas most affected by the rains were in the catchment of the Elbe river comprising the Western part of the Czech Republic, and the mountain ranges surrounding the Czech Republic – the Ore mountains in the North – separating that country from Saxony (East Germany), and the mountains of the Bohemian and Bavarian forest between the Czech Republic, Bavaria and Austria to the South. Starting on August 8 a first extreme rainfall fell on the Moldava catchment – the largest tributary of the Elbe river in the Czech Republic. The catch-

Table 2.4.1 World Bank Strategy for disaster reduction

	Strategic focus		Adjustment in bank instruments and actions		
Development Objectives	Intermediate Goals	Analytical and Advisory Activity	Policy Integration	Project Design and Lending
• Reduce peoples vulnerability to environmental risk, including moderate and extreme natural events • Minimize: • loss of life and livelihood • injuries and disabilities • temporary and permanent dislocation • destruction of social, physical and natural capital	• Raise awareness of the potentially high economic and social returns that investments in vulnerability reduction can yield • Strengthen regional institutions to improve weather forecasting, dissemination, and verification systems • Enable adoption and encourage enforcement of building codes and land use policies • Promote resilience through better management and protection of the natural resource base	• Study the social and economic impacts of natural disasters and assess the vulnerability in countries / sub-regions with a history of natural disasters • Develop a framework for vulnerability assessments, disaster preparedness, and early warning systems • Support the preparation of building codes, siting and land use guidelines • Develop learning programs on planning, predicting and adapting to climate change	• Include disaster prevention and management in policy dialogue • Promote the integration of vulnerability reduction measures in sectoral planning and regulatory reforms • Support the integration of disaster management into regional, national and local land use and development plans and watter resources management policies, strategies and planning	• Support community based ecosystem service initiatives to reduce the impacts of flooding (reforestation, conservation, and restoration of wetlands) • Build and strengthen early warning systems, including community-based systems for effective dissemination of information • Support vulnerability reduction investments, including investments for adaptation to climate change

Fig. 2.4.5 The Elbe catchment in Central Europe

ment of the Danube regions South of the Czech Republic was also strongly affected. A second extreme rainfall event occurred four days later, with two extreme rainfall cells: once again causing floods in the Czech Republic, but the maximum was concentrated on the Ore mountains and the adjacent regions further North. The observed rainfall in parts of the country was more than 350 mm/24 hrs – a rainfall higher than the total in many summers in that area, and as high as the largest daily rainfall ever recorded in Eastern Germany.

The flood: the Elbe catchment was already soaked from the heavy rainfall in July – and the runoff was very high. With the heavy rainfall of August 8th in the Czech Republic, extreme floods occurred – the low lying parts of the city of Prague were partly under water. The flood wave from the first event in the Czech Republic reached its peak on August 9, then moved down the Elbe river into Germany towards Dresden. The second flood wave occurred when the floods from the August 12th/13th rainfall arrived. The superposition of the two events caused enormous damage in the Czech Republic: about 200,000 people had to be evacuated from the flood area, and very high property damage resulted, for example, the subway in Prague was flooded in many parts, and many Czech cities were partially flooded. Losses were estimated to exceed the damage caused by the 1997 flood. In the South, the floods in the Danube caused extensive damage, making the flood one of the highest in comparison with previous floods. In the city of Passau, on the confluence of the Danube and the Inn rivers, the floods flooding the lower parts of the city were the highest observed since 48 years, and further downstream along the Danube in Austria large damage occurred in the cities almost up to Vienna.

The largest damage occurred in the German states of Saxony and Saxony-Anhalt, where the flood waves from the two rainfall events almost coincided. Small cities on the Elbe river (Pirna) and Mulde (Grimma) were caught not well prepared and suffered enormous damages. Among the worst hit was Grimma, which just had been rejuvenated after having been in a very poor state before the 1989 reunification of Germany. It was a lovely city, with all houses freshly remodelled and the streets newly paved, with a lot of engagement by the local citizens. All in ruins! – or at least damaged enormously.

The hardest hit, however, was the city of Dresden, which suffered from an unusual accumulation of contributing causes. On August 13, the second wave of extreme rain centred on the Ore mountains, producing flash floods of unheard magnitude. Small reservoirs, which gave a certain sense of safety to the people downstream, filled up rapidly and overflowed into usually harmless little streams, which became rapid torrents. Among them was the Weisseritz. It had been diverted many years ago from its original bed, and where it once flowed into the city is now the main railway station. The stream broke its diverting dike and smashed into the train station, where the trains drowned in the flash flood. The lower part of the city went under water – including the basements of some of the most spectacular buildings: the Zwinger, (i.e. the castle of the kings of Saxony, which houses the invaluable art collection of the former kings – among the treasures are paintings by Leonardo, Rubens and Rafael – whose wonderful Sistine Madonna alone is worth a trip to Dresden) and the Opera, which is one of the most glorious pieces of the architecture of the 19th century. Fortunately, the concerted effort of many citizens and art lovers made it possible to save all paintings that had been stored in the basements.

And then came the floods in the Elbe river. Floods from the Czech republic and from the or mountains reached the city of Dresden on Tuesday, October 13th and peaked about a week later: the water level was more than two metres higher than the worst flood ever measured – and the problem was aggravated by a very poor forecast: somewhere between August 15th and 16th the water administrator of the city had warned that the level of the flood may reach 8.30 m gauge level – and on August 17th it actually peaked at more than 9.30 m! The Elbe bursting its banks continued the flooding of the city. Of great consequences was the large impediment to transport. In Dresden, the bridges over the Elbe river were closed for days, (as were the bridges over the Moldava in Prague) and the floods over the land destroyed hundreds of kilometres of roads and railways.

From Dresden, the flood moved towards the north: cities like Meissen in Saxony, Dessau at the confluence of Mulde and Elbe, the small city of Mühlberg and a large part of the city of Magdeburg went land under. Dikes broke in many places, and diversions had to be opened, for example in the area of confluence of

Elbe and Havel. The flood diminished so that the cities further to the North were not as badly affected as the ones further South. In the city of Hamburg, where the flood arrived on August 25th, the water administration was not worried at all: compared to the tidal waves that may come in from the North Sea, the flood waves from the river are quite small, and Hamburg is prepared against storm surges with a recurrence interval of about once every 200 years.

The damage everywhere would have been much higher, and many more dikes would have failed if the dikes had not been strengthened through enormous numbers of sand bags, which were filled by everybody: from soldiers to citizens of the affected villages and cities. In the face of this disaster a wonderful solidarity developed, and thousands of people helped – neighbours and helpers of all kinds came to the affected area from all parts of Germany, bringing supplies, equipment, and their own engagement. Together with about 35,000 German soldiers there were groups from other nations – probably more than 1000 soldiers from England, France and the Netherlands volunteered. Nevertheless, in terms of cost, the damage of this flood disaster will be the highest ever incurred in Germany, very preliminary estimates range from 20 to 30 Billion Euro (or US $.) Fortunately, the number of fatalities is very small: probably fewer than 30 persons lost their lives in Germany, and fewer than 15 in the Czech Republic as a direct effect of the flood.

The German Government reacted very smartly (which may have helped chancellor Schröder to win the parliamentary elections in September). It decided to postpone a planned tax cut for 1 year, so that next year something like 8 Billion Euros will be available. The European Community also is helping, and many millions of Euros are coming in from donations by private citizens, companies, clubs, and other organisations. Politicians and financial experts are at work to make plans how to best distribute this large amount of money intelligently and most effectively. The issue is not only that large numbers of households are affected, but also that thousands of small businesses are ruined – businesses, which very often just had started, whose stocks were destroyed, and whose debts are not covered. They were already at the upper limit of what could be financed, because they had to start from scratch after German reunification in 1989. There also is a psychological issue: how to let people feel, that it is worth to start again, after they may have lost everything for which they had worked and toiled during the last ten years. It is hard to assess the damage to the spirit of the people: the first reaction was tremendous solidarity, but there is a let down after the flood is gone, and the realisation sets in on what has to be done to reach the point again that had been reached before the flood. Appropriate encouragement is perhaps the most important support for the people of this part of Eastern Germany.

The flood not only gave occasion to wonderful examples of solidarity and self help, but provided also some insight into what could be done better. The warning system left much to be desired: the rainfall field was too local to be accurately forecasted by the German Weather Service, the warning did not reach people early enough, and much confusion was caused by not having a clear chain of command for the actions that needed to be taken, and the logistics of where to put which group of helpers did not function in many instances. In the Czech Republic and Bavaria, the authorities had learned from 1997 and later disasters but in the Elbe catchment in Germany, the authorities were not prepared for such a disaster. Although a number of studies were available, which indicated the weakness of the flood defences on the Elbe, and although scientists had warned after the flood of 1997 that the events on the Odra river could be repeated on the Elbe, the authorities had largely failed to act. By learning from the experience, we hope that preparedness will be improved.

References

Ahorner L, Rosenhauer W (1993) Seismische Risikoanalyse. In: Plate, EJ et al. (eds) Naturkatastrophen und Katastrophenvorbeugung. Deutsche Forschungsgemein-schaft, Bericht des Wissenschaftlichen Beirats der DFG für die Internatinal Decade for Natural Disaster Reduction (IDNDR), ch 3.3

Ang AH, Tang WH (1984) Probability concepts in engineering planning and design. J. Wiley, New York, vol 2

Azzoni A et al. (1992) The Valpola landslide. Engineering Geology 33:59-70

Betamio de Almeida A, Viseu T (eds) (1997) Dams and safety management at downstream valleys. Balkeema, Rotterdam

Chow VT, Maidment DR, Mays LW (1988) Applied hydrology. International edi-tion, Civil Engineering Series. McGraw-Hill, New York

Dikau R et al. (eds) (1966) Land slide recognition: identification, movement and causes. J. Wiley, New York

Grünewald U et al. (1998) The causes, progression, and consequences of the river Oder floods in summer 1997, including remarks on the existence of risk poten-tial. German IDNDR Series 10e. German IDNDR Committee for Natural Disaster Reduction, Bonn

Grünthal G, Maier-Rosa D, Lenhardt W (1998) Abschätzung der Erdbebenge-fährdung für die D-A-CH Staaten: Deutschland, Österreich und Schweiz. Bautechnik 75:19-33

Homagk P (1996) Hochwasserwarnsystem am Beispiel Baden-Württemberg. Zeitschrift für Geowissenschaften 14:539-546

ISDR (2002) Living with risk: A global review of disaster reduction initiatives. International Strategy for Disaster Reduction. United Nations, Inter-Agency Secretariat, Geneva

Jessen B (1996) Der Flutaktionsplan in Bangladesh. In: Hanisch R, Moamann P (eds) Katastrophen und ihre Bewältigung in den Ländern des Südens. Schriften des Übersee-Instituts Hamburg 33:270-300

Kottegoda NT, Rosso R (1997) Probability, statistics, and reliability for civil and environmental engineers. McGraw-Hill, New York

Ihringer J et al. (1999) Hochwasserabfluss-Wahrscheinlichkeiten in Baden Württemberg. Landesanstalt für Umweltschutz Baden-Württemberg (LfU), Karlsruhe

Madsen H, Rosbjerg D (1997) The partial duration series in regional index-flood modelling. Water Resources Research 33:771-781

Berz G et al (2002) Winter storms in Europe (II): analysis of 1999 losses and loss potentials. Munich Re Group, GeoRisk Research Department, Munich

Plate EJ (1992) Statistik und angewandte Wahrscheinlichkeitslehre für Bauingenieure. Ernst & Sohn, Berlin

Plate EJ (2002) Risk and flood management. Journal of Hydrology 267:2-11

Kowalczak P (1999) Flood 1997 – infrastructure and urban context. In: Bronstert A et al. (eds) Proceedings of the European expert meeting on the Oder flood, May 18. European Commission, Potsdam, pp 99-104

Schmincke HU (2001) Vulkanismus. In: Plate EJ, Merz B (eds) Naturkatastrophen, Auswirkungen, Vorsorge. E. Schweizerbartsche Verlagsbuchhandlung (Nägeli and Obermiller), Stuttgart, ch 2.2

WMO (1999) Comprehensive risk assessment for natural hazards. TD 955. WMO, Geneva

World Bank (2002) Making sustainable commitments: an environment strategy for the World Bank. World Bank, Washington DC

Zschau J, Küppers AN (eds) (2002) Early warning systems for natural disaster reduction. Springer, Berlin

2.5 Global Environmental Change and Human Health

Rainer Sauerborn[1], Franziska Matthies[2]

[1]Department of Tropical Hygiene and Public Health, University of Heidelberg,
Im Neuenheimer Feld 324, 69120 Heidelberg
[2]Tyndall Centre (HQ), School of Environmental Sciences, University of East Anglia, NR4
7TJ Norwich

Health matters

Human health is a prerequisite not only for individual well being, but for development and productivity (Commission on Macroeconomics and Health 2001). Within the debate on global change, it is the primum movens for most people to be concerned about global environmental changes (GEC) and change behaviour or policies to protect their health. It is therefore crucial to understand the pathways how GEC may affect health and to develop policies how societies can protect themselves and/or adapt to changes. Smith et al. (1999) estimate that as much as 25 – 33% of the global burden of disease can be attributed to environmental risk factors. The drivers of global environmental changes encompass climate change, stratospheric ozone depletion, loss of biodiversity, nitrogen loading, over-exploitation of terrestrial and marine ecosystems (including for example soils), depletion and pollution of fresh water supplies and persistent organic pollutants (McMichael and Martens 2002, WBGU 2000). Most of these changes are anticipated to have adverse effects on human health, mediated either directly or indirectly through physical, ecological and social factors and their complex interactions (WHO 2002a).

In contrast to the importance of health within the GEC debate, it is striking how health aspects are treated in a rather marginal way by the academic community. One sign is the under-representation of health in existing global scenario studies, as recently assessed by Huynen and Martens (2002). From 31 selected scenarios, only 14 described health adequately, whereby only 4 considered socio-cultural, economic and ecological factors as driving forces for the development of health (Huynen and Martens 2002).

In this chapter we will argue that there is a particular research need in three areas: (i) health impacts: attribution of effects to GEC, through longitudinal studies and modelling; (ii) adaptation measures: their evidence-based identification and evaluation in different ecological and social settings and finally (iii) development and validation of health surveillance which allows to improve impact models in an

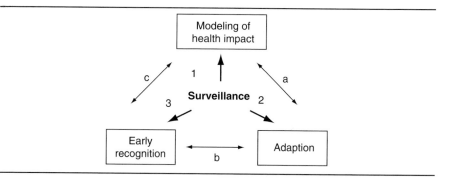

Fig. 2.5.1 Priority issues of health research and their links

iterative process, to evaluate the effect of adaptation and to provide early warning for health effects. These three research issues are closely linked and interdependent as Figure 2.5.1 illustrates. We will elaborate on these links further down in the text. Suffice it here to give two examples to illustrate the point: Health impact modelling should integrate human response (adaptation), early recognition of actual health impact can be fed back into modelling.

The following section 2 gives a brief overview of our current knowledge of health impacts likely to be caused by the different drivers of GEC. We will touch on some specific research questions. Section 3 deals with the challenges of modelling health impact. In section 4 we stress the importance of adaptation processes and point to some large research gaps. We go on to argue for the central role of health surveillance, particularly in developing countries to carry out research within the framework shown in Figure 2.5.1.

Health impact of GEC

This section reviews the state of knowledge on the likely health impact of selected drivers of GEC in turn.

Climate Change

The Intergovernmental Panel on Climate Change (IPCC) predicts a rise in global mean temperature by 1.4 – 5.8°C between 1990 and 2100 and a mean sea level rise of 45 cm by 2100 (Houghton et al., 2001). An increase of extreme events (storms, floods and droughts, El Niño events) will accompany global warming. Warming is expected to be stronger in higher latitudes, especially of the northern hemisphere, and arid and semi-arid regions become drier while precipitation in areas at mid to high latitude increase. The speed with which these changes currently occur and are projected is unprecedented and probably plants, animals and humans may not have enough time to adapt.

Insidiously, health effects will not be new, but gradual increases of existing disease burden will occur. The World Health Report 2002 (WHO 2002a) estimates that in the year 2000 climate change caused 2.4% of worldwide diarrhoea, 6% of malaria in some middle income countries and 7% of dengue fever in industrialised nations. WHO attributes 154,000 deaths (0.3%) and 5.5 million (0.4%) DALYs (disability adjusted life years) to climate change (WHO 2002a).

Many infectious diseases, vector-borne, water- and food-borne are highly sensitive to climate conditions and thus to climate change (Kovats et al. 2001). Higher temperatures and humidity support the development of the respective vector and the pathogen in many cases. Altitudinal and poleward shifts of vectors and pathogens have been observed, e.g. highland malaria in Africa or tick borne encephalitis in Scandinavia (Hay et al. 2001; Kovats et al. 2001, Lindgren and Gustafson 2001, Randolph 2001 a +b). For both examples conclusions need to be drawn with care, since especially disease transmission is influenced by a large number of factors and the consideration of possible confounders is important. An increase in drug resistance of the pathogen, changes in interventions (control methods, drug and diagnostic development), behaviour and human activities (preventive behaviour, vaccinations, awareness) may influence the result. Climate-health relationships and weather-health relationships need to be differentiated through appropriate longitudinal study designs. The time scope of such studies exceeds the funding cycle of conventional research projects and needs to be accommodated through special large scale projects with preferentially international consortia.

The El Niño/Southern Oscillation (ENSO) is a quasi-periodic climate variability on the interannual scale. Climate change is suspected to also influence the frequency and intensity of ENSO events. During ENSO events, changes in sea temperature in the Pacific Ocean, and changes in the atmospheric pressure across the Pacific Basin occur. Extreme weather events, such as floods and droughts, may be accompanying the phenomenon, that occurs every 2-7 years (Kovats 2000). There is increasing evidence that the ENSO phenomenon has influence on the incidence of vector-(e.g. mosquitoes or rodents), water- and food-borne diseases, especially in geographic areas, where disease control is weak (Kovats 2000, Kovats et al. 2001). Changes in disease transmission has been observed especially for malaria, dengue fever, and cholera and other diarrhoeal diseases (Checkley et al. 2000, Hasselmann 2002, Kovats 2000). In Peru for example daily hospital admissions of children due to diarrhoeal disease increased to 200% of the previous rate during the 1997/98 El Niño event (Checkley et al. 2000). Should these observations be transferable to other regions, they could give a hint to what effects global warming might have on the incidence of diarrhoeal diseases. The ENSO phenomenon may be seen as an illustration of the environmental basis of many infectious diseases. The incorporation of climate forecasts into epidemic early warning systems may improve preparedness and adaptive capacity to weather extremes and climatic variability (Kovats 2000). The still conflicting evidence of an impact of climate change on vector-borne diseases shows the need for well designed long-term studies focusing on this specific question and for improved surveillance and monitoring programmes in areas at risk.

Rapidly growing cities, especially in poor countries, are in higher danger of suffering from air pollution and heat waves through the urban heat island effect (Patz and Kovats 2002). Resulting health effects for the exposed population, especially the elderly and people with pre-existing conditions, are increasing incidence of cardio-vascular diseases, stroke and respiratory infections, asthma, obstructive pulmonary diseases and lung cancer, respectively. Air pollution (especially through particulate matter) causes 7.9 million (0.8%) DALYs whereby 42% of this burden is suffered in the West Pacific region and 19% in Southeast Asia (WHO 2002a). A case study during a heat wave in Chicago in 1999 showed that risk factors for heat related morbidity, heat stress disorders and heat stroke were age, pre-existing conditions (such as psychiatric illness) and social isolation (Naughton et al. 2002). Most studies on the direct effects of climate on health are from the North. Our knowledge about direct health effects in various climate conditions of developing countries is scarce and calls for more research.

More evidence for health effects of GEC is needed. To demonstrate causal relationships (attributable risk assessment) we need to think in larger time frames than conventional research projects allow: long data series, in which meteorological data and health data are simultaneously gathered in specific (Kovats et al. 2001), longitudinally observed sites (see chapter health surveillance).

Stratospheric ozone depletion

In 1975 the scientists Crutzen, Molina and Rowland received the Nobel prize for their work first describing the depletion of the stratospheric ozone layer and the role of chlorofluorocarbons (Crutzen 1971, Molina et al. 1974). The increase of terrestrial levels of ultraviolet irradiation increases proportionally more in mid to high latitudes, further away from the equator, where the irradiation has already been high before. Well known adverse health effects of excessive UV exposure are erythema and skin cancers in fair skinned populations. Studies in the United States, Australia and Canada showed that the prevalence of non-melanoma skin cancers has increased more than two fold between the 1960s and 1980s, malignant melanoma incidence has increased by an average of 4% every year since the early 1970s in the United States (www.who.int/peh-uv/healtheffects.htm). As a consequence of increased exposure during the last few decades (due to ozone depletion and behavioural changes) a 10% excess incidence of skin cancer in North-America and Europe is expected by 2050 (Huynen and Martens 2002). UV irradiation also damages the eye in various ways: acute effects are photokeratitis and photoconjunctivitis, both painful but reversible. More serious and long-term damage to the eye is cataract development, for which UVB appears to be a major risk factor and which is the leading cause of blindness in the world (www.who.int/peh-uv/healtheffects.htm). Much less knowledge, but increasing evidence exists on the possible suppressive effects of UV on the human immune system, locally and systemically

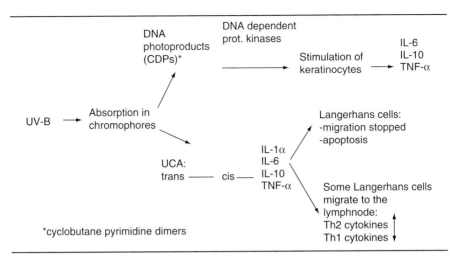

Fig. 2.5.2 Cascades of systemic immune modulation through UV-B irradiation of the skin (drawn after Norval 2001)

(www.who.int/peh-uv/healtheffects.htm; Norval 2001). UV irradiation of the skin triggers a cascade of systemic immune responses, resulting in production of IL-6, IL-10 and TNF-α, apoptosis of Langerhans cells, increase of Th2 cytokines and suppression of Th1 cytokines (Figure 2.5.2).

Local suppression of contact hypersensitivity could contribute to UV irradiation causing skin cancer (Norval 2001). A suppressed immune system increases the risk of infectious diseases, enhances the morbidity and mortality as demonstrated in various animal models of infectious diseases and may even alter the immune response to vaccines. Systemic suppression of the immune system via suppression of Th1 cytokine responses would especially support infections with intracellular micro-organisms. However, most investigations regarding infectious diseases were carried out in rodent models and data on susceptibility to infectious diseases in humans are still missing (Norval 2001). The effects of UV radiation on the immune response to hepatitis B vaccination in humans has been assessed measuring specific and non-specific cellular and humoral parameters in cases (UV exposure) and controls (Sleijffers et al. 2001). In this study, UV exposure prior to vaccination neither influenced lymphocyte stimulation (cellular response) nor antibody titres (humoral response) to the hepatitis B surface antigen, however, contact hypersensitivity responses and natural killer cells were suppressed. The used vaccine induces mainly a Th2 response, leading to an antibody response. A vaccine that triggers particularly a Th1 response would be interesting to investigate, as Th1 responses are more sensitive to UV induced suppression. The possibility of such a major impact of UV exposure on the human immune system as altering responses to vaccination underlines the need of further research and clarification in this area. Open questions (reviewed in Norval 2001) are also whether UV irradiation may be able to induce or activate viral infections. HSV causing so called "cold sores" and HPV and the association with the development of squamous cell carcinomas are known examples. For HIV, controversial effects of UV irradiation could be observed in vitro and in transgenic mice: in UV irradiated cells which contained parts of the viral genome the virus could be activated from latency. However, no effects of sunlight exposure or phototherapy on the progression of disease or immunological parameters in HIV positive patients were observed (Norval 2001). As in most studies the irradiation took place before the exposure to the antigen, so that the induction of the immune response would be affected. It is not known whether also memory mechanisms of the immune system may be altered. Also genetic and other factors responsible for susceptibility or resistance to immune suppression and possible mechanisms of adaptation to UV exposure and compensation of responses to UV exposure still need to be investigated. The methodology for assessing the personal body dose of humans accurately still poses a problem. It is still difficult to exactly quantify the health risks of UV exposure, especially since human behaviour (e.g. dress, shade) has a large influence (de Gruijl et al., in press). Therefore, further investigations of these numerous and diverse aspects could eventually lead to public health messages regarding harmless levels of exposure to sunlight, preventive measures and interventions (Norval 2001).

Loss of biodiversity

Global climate change, stratospheric ozone depletion, chemical pollution, acid rain, introduction of alien species, overhunting of species and the destruction of habitats (such as tropical rain forests) contribute to the loss of biodiversity (Chivian 2001). Plant and animal species are extinguished at high speed all over the world, endangering the stability of essential ecosystems. Experts' predictions estimate that at the current rate, 25% of all living species might be extinct in 50 years time (Chivian 2001). At the same time, vast numbers of valuable chemical compounds and genes are lost before they can be discovered and identified for possible medical use. Medicinal plants are the oldest health care product known. Nowadays, plant parts are either used directly as remedies or their active compounds are used as models for the synthesis of drugs (WHO 1996). In developing countries a large proportion of the population relies on traditional medicine, including herbal remedies, for their primary health care needs. In numerous Asian countries, China, India, Japan or Pakistan, 30-50% of the medicines in use are herbal preparations (WHO 1996). In the European Economic Community about 1.400 herbal drugs were in use in 1991 and complementary medicine in general is in favour of growing numbers of people in the US and Europe. Also, a reduced spectrum of food crops species increases the vulnerability of agricultural systems to plant pests and is a risk for loss of food productivity in longer terms. Thus, loss of biodiversity may mediate increased vulnerability and susceptibility of populations to diseases and infections.

Over-Exploitation of Ecosystems

Among the consequences of over-exploitation of ecosystems through human activities are disturbances of ecosystems (e.g. drought, erosion, fires; Chivian 2002), loss of biodiversity (described above) or soil degradation. Many of those induced changes have direct or indirect effects on human health. Direct effects may arise through contact with contaminated soils, injuries through mud slides or fires. Examples for indirect impacts on human health are reduction of food productivity and transfer of harmful chemicals from soils into drinking water or the food chain (Abrahams 2002). Reduction of food productivity is due to over-exploitation of terrestrial and marine ecosystems, soil degradation, changes in rainfall pattern (droughts) and possible increase of plant pests. Another example for over-exploitation is the situation of the marine ecosystems and the world fisheries (FAO 2001). Many important links in the aquatic food chain have been depleted over the last 50 years of industrialised fishing (Pauly et al. 2000). Disturbance of coastal ecosystems is caused by pollution, degradation of water quality, climate change and mostly by ecological extinction due to overfishing. Ecological extinction of entire trophic levels render the ecosystem even more vulnerable to other natural and anthropogenic disturbances (Jackson et al. 2001, Pauly et al. 2000). In the mid-1990s fish provided more than 50% of the animal protein for the population of 34 countries (FAO 2001). A sustainable management is desperately needed, but under the current economic constraints rather impossible as recent model results show (Kropp et al. 2002). Thus, more knowledge about coastal and marine food webs, including historical, long term accounts of fishing and the effects of legal frameworks and lobbying is needed for successful management and restoration of marine ecosystems (e.g. Jackson 2001). Undernutrition, already among the leading causes of the global burden of disease (27% of children under 5 years of age are underweight, most of them living in Africa and Southeast Asia; WHO 2002a), may increase as a consequence of food scarcity and can in turn render people more susceptible for infectious diseases (diarrhoea, measles, malaria, lower respiratory infections) and may eventually force people to migrate (environmental refugees). The IPCC predicts a reduction in food productivity due to environmental changes for most tropical and sub-tropical regions (McCarthy et al. 2001), most severe, however in sub-Saharan Africa, where land degradation is most prominent and food production decreases (Abrahams 2002). Research is needed regarding future management and conservation of soils, halting and reversing soil degradation, characteristics and pathways of soil contaminants (Abrahams 2002) and sustainable management of marine resources.

Depletion and pollution of freshwater

Water will become a scarce resource over the coming decades, by the year 2050 a quarter of the world population will live in a country that suffers from water scarcity (Tibbetts 2000). Globally, North Africa and the Middle East face the most serious water scarcity today. Irrigation consumes 70% of available fresh water, but growing cities and developing industries will compete more and more for the valuable resource. Thus, water scarcity is also linked to lower food productivity. Receding ground water tables have led to unacceptable levels of arsenic in drinking water for example in India and Bangladesh. Severe water scarcity and droughts have forced people to leave their homes and land (environmental refugees), due to lack of food and water. Policies that encourage wasting water contribute to severe water shortages in many countries. The recently developed water poverty index (WPI; Sullivan 2002) includes resources, access, capacity, use and environment, allowing to rank countries in relation to their provision of water. The results show that rich and richer developing countries score in the top half, with very few exceptions (e.g. Guyana with a high score) and a moderate correlation between the WPI and the Human Development Index (HDI). The WPI may be useful for monitoring and managing valuable water resources globally. However, community level assessments of the water situation are essential for concrete decision making. UNEP recommends increased public private partnership for improved water management.

Unsafe water, hygiene and sanitation cause about 3.7 million deaths (3.1%) and 54.2 million DALYs (3.7%) worldwide, whereby one-third of this burden is carried by the population of Africa and Southeast Asia, and 90% of the deaths occur in children (WHO 2002a). 88% of all diarrhoeal disease in the world are caused by drinking of and contact with unsafe, infected water, lack of water and access to sanitation, as well as poor management of water resources (WHO 2002a). Other diseases linked to unsafe water are schistosomiasis, trachoma, ascariasis, trichuriasis and hookworm infections. Especially in those regions, which already suffer from scarcity of fresh water, the situation is predicted to deteriorate (McCarthy et al. 2001), however, also flooding is a risk factor for water-borne diseases, e.g. cholera outbreaks (WHO 2002b).

More and more people may be forced to migrate by the deterioration of their living environment, lack of resources or conflict over depleted natural resources. Political unrest, social isolation, economic hardship and adverse health effects may in turn be the consequences of migration and rapid urbanisation.

Sparse as the literature on the health impact of single drivers of GEC may be, they leave out interaction between drivers. As an example, increased UV exposure has yet to be determined to have an immunosuppressive effect on humans. How does this influence the impact on infectious diseases, mediated through global warming (vector borne diseases, food-borne diseases) or through freshwater scarcity (lack of personal hygiene). The obvious solution, albeit a daunting one, would be to include all health relevant drivers in a health impact model. We will return to the issue of modelling later. An important objective of health impact assessment must be to identify spatial and temporal differences of exposure.

Adaptation

While mitigation refers to the prevention of global environmental changes, e.g. reducing the extent of global warming by curbing GHG (greenhouse gases) emissions, the term adaptation refers to people coping with or adapting to changes that have already occurred, trying to minimise their effects. Both should be pursued in parallel since there is consensus that even under the most optimistic mitigation scenarios negative health effects are likely to occur.

Adaptation is framed within the concept of vulnerability (Adger 1999) which is defined along three axes:

1) the exposure to the weather or climate-related hazard (includes the character, magnitude and rate of climate variation);
2) the extent to which health, or the natural or social systems on which health outcomes depend, are sensitive to changes in weather and climate (i.e. exposure-response relationship; an example is drinking water contamination associated with heavy rainfall);
3) the adaptive capacity - the ability of institutions, systems and individuals to adjust to potential damages, to take advantage of opportunities, or to cope with the consequences (for example, watershed protection policies, or effective public warning systems for boil-water alerts and beach closings).

Part of the adaptation process will be spontaneous, e.g. when individuals drink more fluids as an unplanned response to a heatwave. Active adaptation which is likely to be more effective requires evidence-based policies and planning (Menne 2002). Ominously, populations likely to be highly exposed lack the resources and the institutions, e.g. health systems, to adapt (WHO 2002a). Therefore, poor countries and the poor in rich countries are likely to be most vulnerable to GEC effects (Patz and Kovats 2002).

The search for scientifically sound adaptation strategies is therefore crucial to help policy-makers identify and implement the most effective and least costly adaptation options.

What factors enhance or impede adaptation, what is the relative contribution of individual, household and system level adaptation strategies? What is the effectiveness of such policies, what are their costs? Can we develop and validate indicators of adaptation? The European collaborative project "Climate change and adaptation strategies for human health in Europe" (cCASHh) currently develops methods for the assessment of population health vulnerability and a framework for adaptation (http://www.euro. who.int/ccashh/). This includes strategies to quantify adaptation processes (Corell et al. 2001) while anthropological methods are needed to understand perceptions, constraints and possible pathways of social adaptation at the individual and household level. The finest possible geographic and "social resolution" is required to identify the most vulnerable populations, i.e. those with greatest exposure and least adaptive capacity. Again, the availability of scientific evidence is inversely related to the problem: where adaptation capacity is lowest, in poor countries, such studies are ominously scarce.

We have dealt with *social* adaptation so far. Another dimension of adaptation which has the potential to influence the health burden of GEC is *biological* adaptation. We will restrict ourselves here to giving a few examples: vectors may change their geographic range, their habitats and feeding behaviour. Reproductive cycles both of vectors and the parasites they harbour may change (shorten to rising mean temperature). Resistance development may be influenced both in vectors and parasites, e.g. the spread of antimalarial drug resistance (Bloland, 2001). Finally the host's defences may adapt to new challenges.

Health surveillance

Monitoring health has three potential benefits in the context of our debate:

1) Improvement of predictive models
   Modelling health impact of GEC has inherent uncertainties and any predictions are bound to have a large margin of error. It is therefore important to assess any GEC –attributable impact and differentiate it from other causes. When health outcomes are measured together with those intermediary processes likely to be influenced by GEC, e.g. entomological inoculation rate, monitoring can contribute to improving mathematical, process-based models (Campbell-Lendrum et al. 2002).

2) Evaluation of adaptation strategies
   Both individual, household level and health systems measures taken to adapt to GEC need ultimately a health outcome assessment to evaluate their effect. If at the same time a set of variables characterising the adaptation strategy and relevant socio-economic data are measured a rich field of analysis of enabling factors and constraints can be assessed at the same time. Costs can be assessed thus allowing to compare adaptation strategies through cost-effectiveness studies. Monitoring health outcomes offers the opportunity to evaluate the effect of adaptation measures, thus closing the triangle between disease modelling, early recognition and adaptation.

3) Early recognition of health impact
   By separating the GEC-attributable signal from confounding factors, such as seasonal fluctuations or resurgence due to drug resistance or a break-down of health service (Mouchet et al. 1998), monitoring of GEC-sensitive health effects helps to guide public health interventions in time and space.

How can health be monitored? The most obvious and widely used approach is by using data from national routine health information systems (Lippeveld et al. 2000). These data are compiled by many international organisations, most importantly through the World Health Organization. However, such data are notoriously imperfect, since in most cases, they are based on passive case-finding, which means that only patients presenting at a formal health service will have the chance to be registered. Particularly in developing countries, health services are rarely utilised. In Burkina Faso, only about 1 in 5 people suffering from an illness, seek formal care at all. This introduces a serious bias, since the utilisers of health services are known to be far from representative of the source population. Incompleteness and irregularity of reporting and transmission and compiling are other sources of bias.

Another method to monitor health is by what is called health surveillance (Sauerborn and Yé 2000). This is an active approach to assessing the health of defined populations. It is a continuous, longitudinal and long-term effort and particularly suited to developing countries where data from routine information systems are incomplete and seriously flawed. It is in these countries that a large part of the GEC-related disease burden is likely to occur. It is therefore crucial to have high-quality health surveillance systems in place.

In the following we will illustrate what we mean by "health surveillance" using the Nouna Health Research Center (www.CRSN-Nouna.org) in Burkina Faso as an example[1]. The Center carries out health surveillance for a total population of 60,000[2] living in 41 villages and the district town of Nouna. Each month, interviewers go from house to house to update census information (1 in Box 2.5.1) and information on vital events having occurred in the preceding month (2 in Box 2.5.1).

Both population and geographical data are geo-referenced and integrated in a health and environmental information system.

The Nouna health surveillance research site is linked with a network of 29 similar sites covering 16 countries in Asia, Africa[3] in very different ecological and economical settings, both in urban and rural areas. The network called INDEPTH (International Network for the Demographic Evaluation of Populations and Their Health) provides the opportunity of a "global" early warning system for changes in the incidence of disease or their temporal and spatial changes (INDEPTH 2002).

Of the 29 health surveillance sites of the INDEPTH network, all sites collect data on items 1 through 4, many collect data on diseases and health care utilisation and geographical information. Only one site (Nouna) systematically collects meteorological data. Great efforts are undertaken to standardise measurements across sites for comparability and to expand data collection to GEC relevant data (see Box 2.5.1).

Another possibility to extend the locally found data from surveillance sites to regional or global scale is provided by remote sensing. Proxies of disease, in the case of malaria, transmission pressure, reflected by the "entomological inoculation rate" EIR, for example, can be recognised through signals from a variety of channels of satellites. This provides the opportunity to validate remote sensing through ground

---

[1] For details we refer the reader to Würthwein et al. (2001)

[2] Typically, health surveillance sites cover entire populations of between 50,000 to 100,000 people, and their ecological environment.

[3] The first Latin American site is being reviewed currently for membership.

Box 2.5.1 Type of data collected by the Nouna health surveillance site

1) **Denominator:**
   enumeration of the entire population at any given month to establish the denominator for vital events and disease: resident population
2) **Demographic data:**
   vital events: death, by cause, age and sex; births; in- and outmigration
3) **Socio-economic data:**
   household composition and size, revenue, expenditure
   Assets Educational attainment
4) **Data on disease incidence and prevalence:**
   disease occurrence (through household morbidity surveys, verification of reported illness with laboratory exams, whenever reasonable and feasible)
5) **Entomological and parasitological data:**
   related to selected, GEC sensitive diseases, such as malaria
6) **Health care utilization:**
   by type of care and level of health service
7) **Geographical data:**
   altitude, surface water, land cover
8) **Meteorological data:**
   precipitation, temperature, wind direction and intensity, humidity, evaporation, radiation

---

data (shown in the inset to the left of Figure 2.5.4) to create regional or global risk maps (Rogers et al. 2002). The MARA (Mapping malaria risk in Africa) collaboration uses a numerical approach based on biological constraints of climate on the development of parasites and the vector to describe and map the distribution of malaria transmission (Craig et al. 1999). The model and the resulting maps can be used for the prediction of the impact of climate change on malaria transmission and under consideration of population, morbidity and mortality data give fundamental information for the planning of control strategies.

The WHO has set up a new global outbreak alert and response network with a computer-driven infrastructure to detect and inform about infectious diseases outbreaks in the world (Heymann et al. 2001). This technical partnership aims at improving global and national preparedness to known risks, new trends (e.g. emerging and re-emerging diseases) and unexpected events and focuses on delivering fast and effective support in the field.

Overarching research challenges

In studying and predicting health impact of GEC and the role of adaptation to it, researchers face a daunting list of problems: a bewildering multitude of health outcomes, the complexity of influences on health, the attribution of adverse health out-

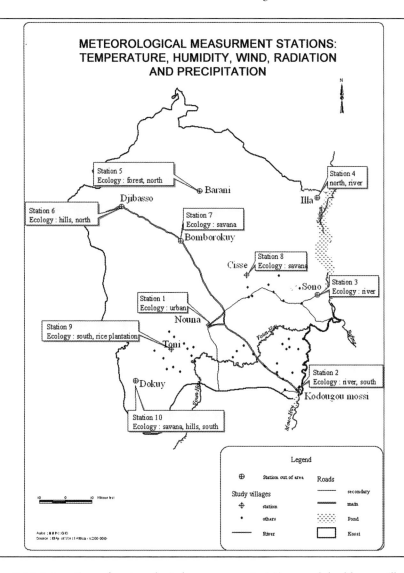

Fig. 2.5.3 Integration of meteorological measurement stations and health surveillance. Source: Nouna Health Research Center 2002

comes to GEC, the fact that most GEC is projected in the future, the long lag times between exposure and outcome. We will discuss them in turn.

1) Definition of an "adverse health outcome"

A multitude of health outcomes is used: incidence and prevalence rates of diseases, chronic and acute diseases often weighted by severity, injuries, deaths

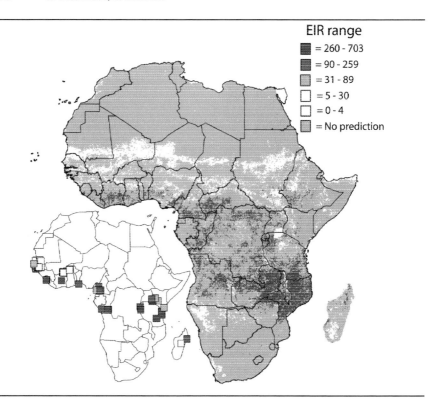

Fig. 2.5.4 Predicting malaria transmission through remote sensing, validated through ground measurement at the sites indicated in the inset on the left. (Rogers et al. 2002)

rates, injury or disability. One and the same disease can produce multiple outcomes (say malaria can cause disability, acute diseases and death) and one facet of GEC can influence different diseases (global warming is likely to affect a host of vector-borne diseases).

The health effects of global environmental change (GEC) in general have traditionally been measured in terms of population mortality as a consequence of such a change. For example, the World Resources Institute 1998-99 estimates that almost 4 million children die each year of acute respiratory infection, linked with indoor and outdoor air pollution, or 1-3 million lives a year are claimed by malaria, a disease closely linked with environmental conditions (WRI 1998). This measure of health effect, however, does not take into account the non-fatal health outcomes (morbidity) and the quality of life associated with it. Considering population mortality, or in some instances even non-fatal health outcomes, separately as health effects of GEC may be useful to appreciate their impact on human health. They are, however, of very little help in

designing prevention and control programmes and making resource allocation decisions because these measures do not provide an acceptable comparability across different conditions (say, diarrhoea, malaria, flood injury and
malnutrition) emerging from the same environmental change (say, climate
change). In other words, measures based on mortality or non-fatal outcomes
alone are inadequate at least for four reasons: a) they do not readily allow for
making comparisons of health effects between two populations or the same
population over time, b) they are unable to quantify health inequalities, c) they
do not provide a common "currency" to measure the magnitude of different
health problems, and d) they cannot be used to analyse the benefits of interventions for use in, say, cost-effectiveness studies. A relatively newer approach
has considered environmental change as a risk-factor of a certain health outcome and measures the health effects as time-based health gaps in the population (World Bank 1993, Ezzati et al. 2002). Expressed in terms of
disability-adjusted-life-years (DALYs; Box 2.5.2) attributable to the risk factor
(environmental change), it offers the possibility of using a common metric
for population health by measuring the difference between actual population
health and some specified norm or goal (see Murray 1994 for technical basis
of DALYs and Ezzati et al. 2002 for a discussion on burden of disease attributable to risk factors).

We propose to use wherever possible, DALYs as a standardised comprehensive measurement for health effects in order to achieve standardised and meaningful impact comparisons across diseases, space and time.

2) Complexity of factors influencing health
A problem that has always haunted studies on the determinants of health is
the sheer complexity of the genetic, biological, social, economic and environmental factors that influence health. Health systems are but one social system
to protect health. The influences of education, agricultural, transport and
information sectors on health and health care are well documented. Isolation
of exposure variables in observational studies (case control and cohort) has
been used to assess their specific contribution. Particularly case control studies have been used to address long time lags between exposure and disease.
Their main limitation, by definition, is their reductionist approach in isolating one factor, ignoring all others, assuming that their influence will ideally
even out between the cases and controls. Another limitation in the validity of
the retrospective exposure assessment and in the inability to capture concurrently other confounding factors diminished their value for health impact
assessment (for further discussion see Woodward 2002).

3) Problem of attribution of GEC variables to adverse health
Rather than anticipating the emergence of dramatic "new environmental diseases", we expect that some already existing health problems will deteriorate in
an insidiously incremental way. This raises the challenge of attribution of
adverse health outcomes to global environmental change. We need more evidence for the processes and pathways through which GEC affects health.

Box 2.5.2 DALYs: time as a measure for health

DALYs or Disability-Adjusted-Life-Years are based on the concept of health gaps. They estimate the impact of a disease or condition in terms of individuals' lost quality of life by measuring the difference between actual population health and some specified norm (perfect health). There are two components in DALYs: a) years of life lost (YLLs) due to premature death, and b) years lived with disability (YLDs) due to a disease.

The "currency" of ill health is therefore the time in years of complete health lost. The loss can be due to premature death or time lived with disability. For the former the maximal life expectancy of humans is used as a normative reference (82 years, female Japanese). Three preferences are used to weight each year of life lost:

Time preference (a discount rate of 3% is used)

Age preference (giving relatively greater weight to productive years of adulthood)

Health state preference; based on studies on the relative importance of diseases for communities, each condition is given a preference weight.

DALYs are expressed formally as

$$\int_{x=a}^{x=a+l} DCxe^{-\beta x}e^{-r(x-a)}dx$$

where:

a = age at onset

D = disability weight (1 for premature death)

C = age weighting correction constant

r = discount rate

a = parameter of the age weighting function

l = duration of disability or time lost due to premature mortality

---

Based on such improved knowledge, we should make careful use of observational longitudinal studies of health and climate and other environmental variables. In addition to information on health and environmental change, such studies should include socio-economic variables (income, assets, expenditures, education etc.) as well as information on adaptation strategies (land use, health care use, risk sharing strategies, migration etc.). Adaptation strategies, if ethically appropriate, could be randomised to population clusters for rigorous testing of their effectiveness. Such studies and understanding of causal chains is needed in order to improve surveillance and control measures and to identify strategies for adaptation.

On a technical note, we propose to use the "population attributable fraction" and the percentage reduction in disease or death that would take place if exposure to risk a factor were reduced to counterfactual distribution, with all other factors remaining the same. The advantage of this method is that it addresses all four weaknesses of traditional measures of health effects as discussed in the preceding paragraph. The methods and results of this approach are described in Ezzati et al. (2002) and the World Health Report (WHO 2002a). The WHO has undertaken DALY estimations for attributable risk of selected environmental factors (e.g. urban air pollution, indoor air pollution, global climate change) in the World Health Report 2002 (WHO 2002a). Estimates for attributable burden of disease for more environmental risk factors is needed, on global, but also on national level for concrete preparedness and adaptation measures.

4) Time lag between exposure and adverse health outcome
This is not a new challenge for epidemiologists. However, the lag period is larger firstly since many of the GEC will occur in the future, and secondly the long time lag between future exposure and later outcome. It should be made clear to funding agencies that cohorts as described above will be maintained for much longer than the usual 3 year funding cycle. 10 years should be the lower bound for any meaningful study design.

5) Capturing geographic and social differences in vulnerability
Vulnerability to adverse health effects, a function of exposure and adaptive capacity, is likely to vary considerably between individuals, households and regions. Naturally, different drivers of GEC will generate different vulnerability profiles. As an example, the health impact of climate change is widely anticipated to be greatest in the poor countries in Sub-Saharan Africa (SSA) and South India. But also within this large geographical frame, within countries and within households, health impacts will differ widely. So will the capacity to adapt to them. It is therefore essential to carry out studies on health impacts with the highest possible geographic and social resolution focussing on the population where we anticipate the health impact to be greatest and adaptation capacity to be lowest.

Modelling: problems and potential solutions

One important prerequisite for being able to plan, to adapt and to take precautions is being able to predict environmental changes and their possible impacts. A powerful tool for the estimation of environmental changes and their possible effects on health is predictive modelling (Martens et al. 2002, WHO 2002a). Deterministic dynamical, and/or probabilistic/statistical approaches are used for these purposes. However, regarding the projection of future developments current models often produce a variety of additional uncertainties rather than a concrete idea of a specific development. For the model projection of malaria and dengue fever transmission under the influence of climate change for example only a few relevant factors are considered (e.g. MIASMA; Martens et al. 1999, Hales et al. 2002). Taking into account a few dynamical aspects, e.g. the natural constraints of parasite development, and using climatological variables from climate change scenarios as predictors for disease transmission, spread of malaria and dengue is predicted for the future. Most of such models integrate socio-economic and behavioural factors or resistance and new drug developments, respectively, however in a restricted manner. These factors are not only difficult to measure, they are even more difficult to project. Neither are precise and localised projections possible for many health outcomes, especially for those that are caused indirectly and by multiple factors. Efforts need to be made to integrate these important risk factors for the transmission and incidence of many health outcomes, such as infectious diseases. Integrated assessment models (IAM; Martens et al. 2002), a rapidly evolving strategy, aim at integrating various factors and disciplines in order to capture the complex interactions between GEC and human health.

Another prevailing challenge in the development of models combining global climate changes and for example parasitic diseases is the combination of the large range of scales, in time (from days to decades) and as well as in space (from μm to km) (Figure 2.5.5). Scaling operations (up and downscaling) in this context may be often doubtful and can provoke a lot of methodological problems. Yet no general technique is in sight to solve the aggregation/disaggregation problem in modelling. Thus, individual solutions need to be identified for each model. However, research is needed to improve these scaling procedures, also for the large ranges encountered when integrating for example parasite development and climatic changes.

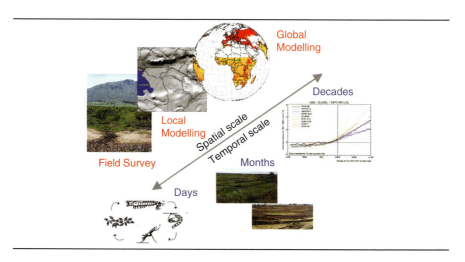

Fig. 2.5.5 Multiple Scales to be considered in modelling of malaria transmission (with the courtesy of P. Martens)

"Regional modules" of models rather than global models would probably provide more concrete input to public health specialists and policy makers for the design of effective interventions and adaptation measures to be implemented locally.

Priority areas of research are therefore the integration of natural, environmental, socio-cultural, socio-economic, pharmaceutical, technological and behavioural factors and the question of harmonising scales. Each of these desirable targets is probably reached individually by solving concrete problems for individual models, such as the question of scale.

In the field of GEC, driven by complex and dynamic ecological, biophysical and social processes, many factors, cause-effect relationships and even the probability of occurrence remain uncertain and cannot be expressed in defined variables (O'Riordan and McMichael 2002). This uncertainty is carried into the estimations of impacts and outcomes and leaves great uncertainty in the results and predictions of models. O'Riordan and McMichael (2002) express the need to "look at both the degree of certainty about cause and effect (outcome) and the probability of occurrence (likelihood and distribution). That dimension of certainty, or lack of it, is integral to both risk estimation and precautionary policies in the modern age."

Ultimately, research results from vulnerability and adaptation, surveillance and control and modelling projects should inform each other and provide input in order to develop intervention strategies and control measures to minimise the impact of GEC on human health.

References

Abrahams PW (2002) Soils: their implications to human health. The Science of the Total Environment 291:1-32

Adger N (1999) Social vulnerability to climate change and extremes in coastal Vietnam. World Development 27:249-269

Bloland PB (2001) Drug resistance in malaria. World Health Organisation, Geneva

Campbell-Lendrum DH et al. (2002) Monitoring the health impacts of global climate change. In: Martens P and McMichael AJ (eds) Environmental change, climate and health – issues and methods. Cambridge University Press, Cambridge

Checkley W et al. (2000) Effects of El Niño and ambient temperature on hospital admissions for diarrhoeal diseases in Peruvian children. The Lancet 355:442-450

Chivian E (2001) Environment and health: 7. Species loss and ecosystem disruption – the implications for human health. JAMC 164:66-69

Chivian E (ed) (2002) Biodiversity: its importance to human health. Interim executive summary. Center for Health and the Global Environment, Harvard Medical School, Cambridge

Commission on Macroeconomics and Health (2001) Macroeconomics and Health: investing in health for economic development. Report. WHO, Geneva

Corell R, Cramer W, Schellnhuber, HJ (2001) Methods and models of vulnerability research, analysis and assessment. Symposium at Potsdam Sustainability Days, Potsdam

Craig MH, Snow RW, le Sueur D (1999) A climate-based distribution model of malaria transmission in Sub-Saharan Africa. Parasitology Today 15:105-111

Crutzen PJ (1971) Ozone production rates in an oxygen-hydrogen-nitrogen oxide atmosphere. Journal of Geophysical Research 76:7311-7327

Ezzati M et al., Comparative Risk Assessment Collaborating Group (2002) Selected major risk factors and global and regional burden of disease. The Lancet 360:1347-1360

FAO (2001) The state of world fisheries and aquaculture 2000. Food and Agricultural Organization of the United Nations, Rome

Gruijl de FR et al. (2003) Health effects from stratospheric ozone depletion and interactions with climate change. Photochem Photobiol Sci 2003(2):16-28

Hales S et al. (2002) Potential effect of population and climate changes on global distribution of dengue fever: an empirical model. The Lancet 360:830-834

Hasselmann K (2002): Is climate predictable? In: Bunde A, Kropp J, Schellnhuber HJ (eds) The science of disasters: climate disruptions, heart attacks, and market crashes. Springer, Berlin, pp 141-169

Hay SI et al. (2001) Malaria early warning in Kenya. Trends Parasitol 17:95-99

Houghton JT et al. (eds) (2001) Climate change 2001: the scientific basis. Working group I contribution to the Third Assessment Report of the IPCC. Cambridge University Press, Cambridge

Huynen M, Martens P (2002) Future health. The health dimension in global scenarios. ICIS, Maastricht University, Maastricht

INDEPTH (2002) Population, health and survival at INDEPTH sites. In: Population and health in developing countries. International Development Research Centre, Ottawa, vol 1

Jackson JBC et al. (2001) Historical overfishing and the recent collapse of coastal ecosystems. Science 293:629-638

Kovats RS (2000) El Niño and human health. Bulletin of the WHO 78:1127-1135

Kovats RS et al. (2001) Early effects of climate change: do they include changes in vector-borne disease? Phil Trans R Soc Lond B Biol Sci 356:1057-1068

Kropp J, Eisenack K, Zickfeld K (2002) Assessment and management of critical events: the breakdown of marine fisheries and the North Atlantic thermohaline circulation. In: Bunde A, Kropp J, Schellnhuber HJ (eds) The science of disasters: climate disruptions, heart attacks, and market crashes. Springer, Berlin, pp 141-169

Lippeveld T, Sauerborn R, Bodart C (eds) (2000) Design and implementation of health information systems. World Health Organisation, Geneva

Lindgren E, Gustafson R (2001) Tick-borne encephalitis in Sweden and climate change. The Lancet 358:1731-1732

Martens P et al. (1999) Climate change and future populations at risk of malaria. Global Environmental Change 9:89-107

Martens P, Rotmans J, Rothman DS (2002) Integrated assessment modelling of human health impacts. In: Martens P, McMichael A (eds) Environmental change, climate and health. Issues and research methods. Cambridge University Press, Cambridge

McCarthy JJ et al. (eds) (2001) Climate change 2001: impacts, adaptation and vulnerability. Working group II contribution to the Third Assessment Report of the IPCC. Cambridge University Press, Cambridge

McMichael AJ, Martens P (2002) Global environmental changes: anticipating and assessing risks to health. In: Martens P, McMichael AJ (eds) Environmental change, climate and health. Issues and research methods. Cambridge University Press, Cambridge

Menne B. (2002) Vulnerability and adaptation to global environmental change. Paper presented at a seminar on "Global Environmental Change and Health", April 30, Heidelberg University

Molina MJ, Rowland FS (1974) Stratospheric sink for chlorofluoromethanes, chlorine atom catalysed destruction of ozone. Nature 249:810-812

Mouchet J et al. (1998) Evolution of malaria in Africa for the past 40 years: impact of climatic and human factors. Journal of the American Mosquito Control Association 14: 121-130

Murray CJ (1994) Quantifying the burden of disease: the technical basis for disability-adjusted life years. Bull World Health Organ 72:429-445

Naughton MP et al. (2002) Heat-related mortality during a 1999 heat wave in Chicago. Am J Prev Med 22:221-227

Norval M (2001) Effects of solar radiation on the human immune system. J Photochem Photobiol B 63:28-40

O'Riordan T, McMichael A (2002) Dealing with scientific uncertainties. In: Martens P, McMichael AJ (eds) Environmental change, climate and health. Issues and research methods. Cambridge University Press, Cambridge

Patz JA, Kovats SR (2002) Hotspots in climate change and human health. BMJ 325:1094-1098

Pauly D et al. (2000) Fishing down aquatic food webs. American Scientist 88:46

Randolph S (2001a) Tick-borne encephalitis in Europe. The Lancet 358:16-18

Randolph SE (2001b) The shifting landscape of tick-borne zoonoses: tick-borne encephalitis and lyme borreliosis in Europe. Phil Trans R Soc Lond B Biol Sci 356:1045-1056

Rogers DJ et al. (2002) Satellite imagery in the study and forecast of malaria. Nature 415:710-715

Sauerborn R, Yé Y (2000) Demographic surveillance systems: a tool for the analysis of patterns of tropical diseases. New challenges in tropical medicine and parasitology, 18 – 22 September, Oxford

Sleijffers A et al. (2001) Influence of ultraviolet B exposure on immune responses following hepatitis B vaccination in human volunteers. J Invest Dermatol 117:1144-1150

Smith KR, Corvalan CF, Kjellstrom T (1999) How much global ill health is attributable to environmental factors? Epidemiology 10:573-584

Sullivan C (2002) Calculating a water poverty index. World Development 30: 1195-1210

Tibbetts J (2000) Water world 2000. Environ Health Perspect 108:A69-A72

WBGU (2000) Welt im Wandel. Neue Strukturen globaler Umweltpolitik. Springer, Berlin/Heidelberg/New York

WHO, cCASHh project web page. http://www.euro.who.int/ccashh/

WHO, Intersun – The Global UV Project. http://www.who.int/peh-uv/healtheffects.htm

WHO (1996) Traditional Medicine. Fact Sheet 134

WHO (2002a) Reducing risks, promoting healthy life. The World Health Report 2002. World Health Organization, Geneva

WHO (2002b) Flooding: health effects and preventive measures. Fact Sheet 05/02

Woodward A (2002) Epidemiology, environmental health and global change. In: Martens P, McMichael AJ (eds) Environmental change, climate and health. Issues and research methods. Cambridge University Press, Cambridge

World Bank (1993) Investing in health. World Development Report 1993. Oxford University Press, New York

WRI (1998) World resources: environmental change and human health. World Resources Institute, Oxford University Press, Oxford. http://wri.igc.org/wri/wr-98-99

Würthwein R et al. (2001) Measuring the local burden of disease: analysis of years life lost in Sub-Saharan Africa. Int J Epidem 30:501-508

2.6 Urban Disasters as Indicators of Global Environmental Change: Assessing Functional Varieties of Vulnerability[1]

James K. Mitchell

Department of Geography, Rutgers University, Piscataway, NJ 08854-4085

Introduction

Cities are where a majority of the world's population now lives and they are likely to become the overwhelmingly dominant habitat of mankind in the near future. Anything that threatens the viability of cities also threatens the survival of the human species. For this reason, it is important that students of global environmental change be able to assess the disaster potential of cities and the means by which unwelcome hazards might be mitigated. (Mitchell 1998, 1999) Such a task raises difficult methodological and philosophical problems for global change analysts. Not only are cities the most intensely human-constructed of places, where dramatic departures from natural baselines are routine; the same processes that contribute to negative environmental changes are often welcomed as agents of desired sociopolitical, economic and cultural shifts. These fundamental contradictions deeply complicate the process of urban hazard analysis.

How best can researchers assess the environmental disaster potential of cities? Is it possible to devise a methodology that does justice to their societal as well as their environmental complexities; that takes account of their human functions as beacons of hope and crucibles of creativity while also addressing their social pathologies and inequities, the environmental burdens they impose, and the potential for catastrophic losses they possess. A comprehensive treatment of these questions would exceed the space available herein. So, for the sake of convenience and simplification the topic of climate change in large cities of More Developed Countries (MDCs) will serve as a surrogate for other kinds of global environmental change in other places. This emphasis has two additional advantages. First, it focuses

---

[1] Based on a presentation to the Symposium on Disaster Reduction and Global Environmental Change, Federal Foreign Office, Berlin, Germany, June 20-21, 2002.

attention on affluent cities that are the disproportionate drivers of many global risks but whose hazard vulnerabilities have been largely overlooked by researchers.[2] Second, it gives access to literatures about cities from the social sciences and humanities that employ plural models of urban processes and divergent interpretations of urban futures. The existence of such models suggests that it will be difficult to integrate research findings about global environmental change into a single interpretive framework of Nature-Society relations. It may be time to explore alternative theoretical formulations that are capable of incorporating qualitatively different kinds of discourse about global change.

Background

Modern scientific research has sought to explain changes that take place in the physical world as outcomes of natural processes. Now, as a result of studies carried out during the last three to four decades, these views have been revised. It is widely believed that humans are capable of acting as environmental modifiers on a scale that bears comparison with natural processes and that mankind has begun to fundamentally transform the global climate as well as other environmental systems. What were once labeled "natural" hazards and "natural" disasters are now perceived as the joint products of nature and society. (Box 2.6.1) Risks (broadly equatable with extremes of natural and human modified systems) interact with vulnerabilities (broadly equatable with potentials for loss in human and built environments) to create hazards (i.e. threats that fall within the coping capacity of society) or disasters (i.e. threats that exceed the coping capability of society).

Box 2.6.1 Hazards and disasters as interactive systems

(HAZARD/DISASTER) = (RISK) + (VULNERABILITY)

HAZARDS fall within the coping capability of society
DISASTERS exceed the coping capability of society

(RISK) = Probability of experiencing a threatening event
(VULNERABILITY) = (EXPOSURE) + (RESISTANCE) + (RESILIENCE)

EXPOSURE: At-risk population and property
RESISTANCE: Measures taken to prevent, avoid or reduce loss
RESILIENCE: Ability to recover a prior state or achieve a desired state

---

[2] The vast bulk of global change research on cities focuses on large, rapidly-growing cities of Less Developed Countries. Most of the relatively small literature on cities of affluent societies addresses issues of urban metabolism and the human modification of urban ecosystems. (Wolman 1965, White 1994, Platt, Rowntree and Muick 1994, Inoguchi, Newman and Paoletto 1999).

To some extent these changed interpretations are part of a continuing intellectual debate about modes of scientific explanation. In this respect, some scholars consider them to be victories for social constructionist interpretations of nature (Eder 1996, Demerrit 2002) at the very least they have blurred traditional distinctions between categories of "natural" and "human-caused". Whatever the merits of the debate, throughout it cities – especially large affluent cities in More Developed Countries (MDCs) – have provided the most convincing examples of human capacities for taming and reconstructing nature to suit our needs and our desires. Cities are both the ultimate human-constructed habitats and the places that are best insulated from damaging natural extremes.

But now we are entering a new phase in the relationship between humanity and nature. Until 15 years ago most disasters in affluent countries occurred in rural districts, small towns and suburbs rather than in large cities and metropolitan areas. Beginning in the late 1980s that pattern began to change and unprecedently expensive disasters – as well as some that are more deadly than heretofore – have begun to occur in what had formerly been considered safe urban areas of North America, Japan and Western Europe (Table 2.6.1) This unwelcome development raises a new problem of interpretation that affects the use of disasters as environmental indicators and indirectly raises broader questions about the meanings of urban vulnerability.

Table 2.6.1 Great urban disasters in MDCs: 1989-2002

Date event	Urban area	Dead	Costs US B (10^9)
1989 Earthquake	San Francisco Bay	62	8
1991 Wildfire	Oakland, CA	25	2
1992 Hurricane	Dade County, F	38	30
1994 Earthquake	Northridge, CA	57	40
1995 Earthquake	Kobe, Japan	6,425	147+
2001 Storm	Houston, TX	22	5
2001 Terrorism	New York, NY	2,837	83

Cost data include (documented) insured losses and (estimated) uninsured losses

Disasters as Environmental Indicators

It has long been argued that an increase in the number and severity of storms, floods, droughts and other natural risks will likely be among the earliest and most burdensome consequences of anthropogenically-forced climate change. (Pielke and Pielkie 1997, Intergovernmental Panel on Climate Change 2001, p. 92). By implication, the world may be facing a more disastrous future, with larger human death tolls, bigger bills for economic losses, increased disruption of communities and ecosystems as well as heavier costs for protection. This is a message offered by many global change specialists and commentators. And it is a message whose propagators frequently search for confirmatory indicators. Just as climate scientists interrogate each passing year's temperature and precipitation records for evidence that a warmer world has truly arrived, so too do students of natural hazard scan the latest tabulations of loss statistics looking for deteriorating trends of disaster. As absence of bigger natural extremes, formerly dependable measures against hazard can fail because Table 2.6.1 suggests, these may already be evident; but their meaning is far from certain.

Worsening physical risks do not invariably produce larger disasters. Connections between risk and disaster are usually indirect, being filtered through conditions of social vulnerability that are themselves joint products of exposure, resistance and resilience. As a result, the incidence of disaster may rise or fall under the aggregate effect of many factors that separately are moving at different rates, in different directions. For example, improved programs of hazard management can offer adequate protection even if physical risks are deteriorating or if more people are relocating into harm's way or if communities have lost some of their capacity for rapid recovery after disasters occur. Likewise, even in the event of neglect or if behavioural and demographic shifts put more people and investments at risk. In view of these kinds of complications, attempts to correlate disaster trends with climate changes (or with changes in the riskiness of other natural systems) are frequently misleading.

The difficulties of linking climatic fluctuations with a potential for increased disasters are greatest in urban areas. Cities are characterised by enormous complexity and dynamism; they are constantly remaking themselves in response to broad social forces as well as the wishes of powerful social agents. Often the forces that most change the disaster potential of cities are not connected with physical threats like those triggered by climate shifts. They are a product of social, political, economic, technological and cultural vulnerabilities that have developed along with the cities themselves. Finally, it is not axiomatic that city governments and other institutions will pay particular attention to environmental risks and vulnerabilities. Such concerns must compete with many other considerations on a constantly changing public agenda. This does not mean that global environmental change is - or will be - unimportant in cities, just that it is one of many factors that shape urban habitats. The Intergovernmental Panel on Climate Change (IPCC) has recognised as much in its Third Assessment:

> "There are multiple pressures on human settlements that interact
> with climate change. ... these other effects are more important in
>   the short run; climate is a potential player in the long run."
> (IPCC Third Assessment Report, Working Group II, 2001, p. 409)

It is sufficient to recall the recent history of a city like Berlin to realise the vast importance of changes in the ideological, political, demographic, economic, architectural and infrastructural dimensions of urban living. Since the fall of the Berlin Wall in 1989 such changes have probably done more to alter the mix of future environmental risks, levels of exposure and capabilities for coping in Berlin than have long run anthropogenic greenhouse effects or other human alterations of natural processes. For example, the merger of East German and West German political-economic systems that exerted different kinds of demands on regional ecosystems is likely to have far-reaching consequences for – among others - the city's water budget, its high – standing water table, and its prospects for constructing the kind of massive underground infrastructure projects that are deemed necessary for Berlin to assume a hoped-for role as the geographical pivot of an eastward-expanding European Union. Because of these societal shifts, one cannot say with any certainty that deteriorating climate changes in Europe or Germany will bring more or worse "natural" disasters in Berlin – though those possibilities cannot be ruled out either.

An even more striking example of how human-driven contingencies can affect the assessment of urban hazards has been the change in New York City's public stance toward environmental risks after the events of September 11, 2001. Broadly speaking, as reflected in policy statements, budget requests and administrative reorganisation plans of federal and city governments, there has been an enormous turnaround in both the means and the ends of urban hazard management (Armstrong 2002). The importance of emergency preparedness and emergency response have increased mightily and there has been a corresponding reduction in the amount of attention that is paid to underlying social and environmental causes of urban disasters – natural as well as human-made. Whether New York or the United States will be better able to cope with a wide range of such threats in the future is an open question (Mitchell 2003).

Approaches to assessing Urban Hazard Impacts of Climate Change

Global climate change is currently assessed mostly by means of General Circulation Models (GCMs). These involve simulating future climates by manipulating mathematical equations that summarise major atmospheric processes. Vast amounts of empirical data and enormous computational power are required so GCMs are heavily reliant on recently developed global monitoring systems and advanced electronic computers.

Urban areas are a challenge to modelers of climate change, in part because the existing GCMs do not scale down very well to the level of individual cities. The GCMs provide generalised temperature and moisture data for large regional grid-squares whereas cities are relatively small and unevenly developed places. There is another branch of atmospheric modeling that deals with urban microclimatic phenomena such as heat islands, local precipitation anomalies, metropolitan patterns of air pollution and Venturi effects of wind flows among highrise buildings. A wide gap separates these two endeavors both in terms of the topics and time horizons that are addressed. But even if GCMs and the tools of urban microclimate research could be modified to provide better information about city-scale climate changes, it is not at all clear that a shift to greater urban climate variability would result in more or worse disasters.

One of the most wide ranging and well-executed efforts to assess climate change effects on urban disaster potential has been conducted as part of the Metropolitan East Assessment Project of the U.S. National Climate Assessment Program (2002). As the project reports show, it isn't easy to design or execute an urban climate change hazard assessment. Of the three sets of variables that are subject to change – climate, disasters, and urbanisation - this project focused strongly on just the first. It looked at changes in sea level and flood potential that were thought likely to occur during the next 50 years but it applied these data to the New York Metropolitan Area as it presently exists – not taking account of how hazard management systems might change or the community might be altered over that time span. Despite a prodigious amount of analytical work the project investigators concluded that the results were probably subject to large errors because much of the data necessary to conduct climate change research at the metropolitan scale were never collected or properly compiled or were incomplete (see Box 2.6.2). While the task of redressing these deficiencies is not necessarily impossible, it is certainly daunting.

It is safe to say that even with the best available scientific techniques there will still remain large uncertainties about the physical dimensions of urban climate change. This point is significant, not so much because it underscores the limitations of natural science and engineering approaches to analysis but because it counters an all-too-popular notion that it is the human dimensions of climate change, hazards and urbanisation that are "soft" and unknowable or unpredictable. Uncertainties abound in both the natural and social aspects of climate change but they should not be grounds for failing to do the best that is possible with the information and ideas that are available.

Box 2.6.2 What's missing in New York?: Gaps in urban hazard data and analyses.

- Complete catalog of historic storms, coastal flood heights/extents, and associated damages and losses
- High-resolution model of near-shore topography
- Climate models that account for sea level rise as well as variations of storm frequency and intensity
- Inventories of major infrastructure systems with exact location and associated dollar values
- Infrastructure component and network fragilities with respect to storm surge, flooding and wind hazards
- GIS-based computer algorithm for computing losses both probabilistically and for individual events

Source: Metropolitan East Assessment Project

What is needed is an analysis that simultaneously addresses changes that affect all three aspects of the urban climate change problematic – climate risks, disaster vulnerabilities and urbanisation trends. For that it will be necessary to combine something like the Metropolitan East Coast Assessment with a vulnerability forecasting analysis that is informed by a wide array of urbanisation models. Although the direction of previous research has been towards collapsing all 3 sets of variables into a common model, it is not clear that a single model that combines global change, human adjustment to hazard and urban change is yet possible, or if it will ever be possible. One reason for this uncertainty is that the models that evolved to address climate, disasters and cities emerged in different contexts at different times using different types of data and subject to different sets of assumptions. They are now so diverse that it is difficult to see how they might be fitted into any single master framework. Instead, it may be worthwhile to seek other ways of understanding and adjusting the relationship between society and nature in cities.

Metaphor-Based Models of Cities

At the heart of the urban climate change problematic is an awkward reality; over the last two centuries no human institution has been subject to so many different types of analysis and commentary, nor given rise to so many different interpretations, nor engaged so much of the collective energy of critics, reformers and revolutionaries, nor attracted so many prescriptions for the future, as the city. This means that if analysts are to assess the vulnerability of cities to environmental change they must find a way of linking the different conceptions of cities and the different envisaged urban futures with what is known about the risks of climate change. To facilitate this goal it is useful to think of the inhabitants of large cities as being engaged in many different "conversations" about society, the built environment and the natural systems in which they are embedded. These conversations are characterised by shifting frames of reference among diverse urban interest groups. They are perhaps better understood metaphorically than empirically. The range of possible metaphors about cities is probably infinite but for the purposes of this analysis six examples will suffice. These include metaphors of the city as:

> **Box 2.6.3** Six Metaphors of Cities
>
> • Machines
> • Organisms
> • Learning systems
> • Regulated territories
> • Performance spaces
> • Muses

(1) machine; (2) organism; (3) learning system; (4) regulated territory; (5) performance space; and (6) muse. Each metaphor has its own substantive focus and preferred set of policy mechanisms and – most importantly in this context – each is subject to its own kind of vulnerability.

Cities as Machines

Most schemes of urban management owe a debt to the metaphor of cities as machines. This notion is rooted in Western society's romance with technology-driven change, from the Renaissance to the Industrial Revolution and beyond. Machine models focus on the urban fabric and the organisation of materials that makes possible the execution of characteristic urban functions such as the assembly, sheltering and distribution of goods, services or people. When used as prescriptive tools machine models tend to employ principles of optimisation or perfectibility. These suggest that it is possible to redesign the urban machine so that it is capable of operating more effectively under an increasingly wide range of constraints.

How is vulnerability manifest for this metaphorical model? The vulnerability of the city as a machine is signaled by the failure of its components to perform according to design specifications. Indicators of failure include disruptions, breakdowns, sub-optimal performance levels and other changes that impair the delivery of services. This kind of vulnerability can readily be measured and such assessments constitute the most common kind of urban vulnerability analysis. Evaluations of building and infrastructure responses to environmental threats carried out in connection with the Federal Emergency Management Agency's HAZUS Mapping program, are good examples. (e.g. New York City Consortium for Earthquake Loss Mitigation 2002 http://www.nycem.org/default.asp). Despite their lack of attention to most human dimensions of vulnerability, assessments of this type – that combine information about forces that impact on cities (e.g. climate change risks) and information about the physical vulnerability of structures and infrastructures – are often regarded as state-of-the-art examples of integrated hazard analysis.

Cities as Organisms

The metaphor of cities as organisms also has a long lineage in western thinking, from Biblical images of Paradise, through classical Islamic city planning schemes, to Leonardo da Vinci, the Garden Cities movement, social Darwinism and recent advocates of ecological cities. In this case the focus is on interactions between society and nature, especially as these are expressed in the reproductive and metabolic processes of cities. Concepts of thresholds and boundary conditions that separate different states of punctuated equilibrium are important aids to thinking about organic urban change. Under the rubrics of sustainable development the city as organism should be designed to ensure the long-run welfare of its residents and of areas within its ecological footprint, both in light of prevailing and foreseeable environmental risks or opportunities. The means by which these goals are achieved include existing adjustments to environmental stresses and the expansion of capacities for further flexible responses.

In this context, urban vulnerability is a matter of threats to life-support systems rather than to technologies; the impacts of environmental change on common property resources – like the atmosphere, the oceans and non-human life forms - are just as significant as the impacts on immediate human safety or on the security of buildings and infrastructures. In principle, information about global change risks should be compatible with the organic model of cities but the integration of these perspectives is still a work in progress. For example, techniques for the valuation and synthesis of seemingly incommensurate human and non-human components of the living environment are still under development. Although a concept of sustainable hazards management has been promulgated (Mileti 1998), the apparent contradictions of safety and sustainability as criteria for urban management have yet to be explored in detail.

Cities as Learning Systems

Organisms are one kind of adaptive system; formal learning systems are another. There are several advantages to casting the city as a learning system. First, the concept of learning systems imparts forward momentum to the analysis of global change by going beyond equilibrium notions of adaptation to capture the fundamental transformations of nature that are now generated by human activities. (Cronon) Second, learning about dynamic urban landscapes, hazards and subcultures is a practical task for migrants who are flocking to cities in large numbers, especially in LDCs, and for the millions more who visit as commuters, tourists and other temporary residents. To the extent that global change analysts understand such learning they will be in a better position to affect the future ecology of urban risks and hazards. Third, learning foregrounds the importance of information and it is the acquisition, transmission, interpretation and use of electronic information that underpins the economic and political hegemony of cities in the post-industrial era; for many MDC cities the ICT (Information and Communications Technologies) sector is the single most important contributor to urban employment and economic productivity. Fourth, if a common global culture is developing, it is to a large extent based on the emergence of shared meanings and new institutional relationships among residents of urban areas. Since these areas are continually remaking themselves, learning and forgetting take on particular importance, especially with respect to matters of awareness, representation and action that are basic to the creation of new societal formations. Finally, a learning system model of urbanisation holds open the door to a dimly visible future in which virtual information, and the vast increase in indirect experiences of environments that it makes possible, may be the chief restructuring agents of urban forms and functions.

Viewed from a learning perspective, urban vulnerability involves anything that would impair learning about cities and global change by their inhabitants, managers and leaders. This might include reductions in the gathering or publication of information about environmental conditions; failure to undertake post-hoc evaluations of urban management policies or to act on the findings, demotion of science and reasoned inquiry as bases for the formulation and execution of public policy, and similar actions. Moreover a learning systems model of cities would imply consideration of vulnerabilities that might prevent the emergence of new societal forms and structures as well as those that threaten existing organisms, institutions or infrastructures. Related public policy tools include social monitoring and modulation of emergent behaviours and incentives for innovations that would accommodate, take advantage of – or otherwise respond to – such behaviours.

To a significant degree the preceding metaphors have been operationalised as sophisticated formal models of cities. The remaining three metaphors have not yet given rise to similar formalisations; they remain as embryonic notions. However, this does not mean that students of global environmental change can ignore them.

Cities as Regulated Territories

Cites are good examples of regulated territories. The notion of regulation is twofold. First, it refers to the means by which order is created in urban space and second it carries a more refined connotation in relation to recent theories of regulation that are associated with urban geography and urban political ecology.

Regulation makes it possible for large numbers of people to share the same urban territory without continuous rancorous conflicts about sovereignty, jurisdiction, autonomy and other attributes of legitimacy. This is achieved by creating order around certain norms of authority that are either handed down in taken-for-granted traditions and institutional practices or are negotiated among contemporary interest groups. Among others, these norms include customs, rules, standards, guidelines and laws. Few studies have examined the extent to which urban regulations are implicated in the creation or exacerbation of environmental change problems; issues of regulation are usually viewed from the perspective of measures that might facilitate hazard mitigation if enacted in response to ongoing changes. Nor has the vulnerability of different regulatory systems to environmental change stressors attracted much attention from global change analysts.

Whereas, in the past it was common for urban researchers to take account of existing governmental regimes and changes in their regulatory **policies** that affect cities, it is now evident that entire urban regulatory **systems** can change dramatically and rapidly with far-reaching results. For example, during the 20th century many European cities have been subject to several contrasting regulatory systems that accompanied changes in national or transnational sociopolitical ideologies. These include: imperialism, nationalism, liberalism, fascism, socialism, communism, welfare-statism and neoliberalism. Different systems have attempted to reorganise basic relationships among individuals, economies, polities and societies, thereby altering the process of urbanisation and bequeathing legacies that are likely to persist for decades to centuries thereafter. The recent contrast between city government policies that support (ostensibly risk-taking) entrepreneurial capitalism in London and (supposedly risk-averse) community-stabilisation in Paris is one example. The most recent round of European regulatory shifts involves readjustments in response to the collapse of the Soviet Union, insurgent neoliberal philosophies of capitalism, institutional rearrangements in support of an expanding European Union; and a nascent cultural shift-to-risk in the wake of environmental disasters like Chernobyl and international terrorism incidents like those of September 11, 2001. Inasmuch as changes of regulatory systems sometimes alter the limits of permissible behaviour and often introduce new patterns of rewards or penalties, their influence on urban vulnerability is potentially very large.

One further aspect of a regulatory perspective on urban vulnerability is worth noting. This is the fact that **crises** of varying kinds (usually economic or political) play important roles in regulation theories. For some theorists the city is a predominantly stable setting that is affected by occasional crises; for others it is subject to more frequent crises and for others it is a venue marked by a continuing series of crises. Some of the crises pose challenges to existing regulatory regimes; others threaten regulatory systems; and yet others provoke shifts from one regulatory system to another. Since climate change is customarily viewed as a crisis or a potential crisis, it will be important to consider if – or how – it differs from other crises that beset large cities.

Cities as Performance Spaces

Beginning with the work of the sociologist Goffman (1959) and continuing in more recent contributions to anthropology (Turner 1986, Rapport 1998), geography (Lefebvre 1991) and postmodernist thought (Yeatman 1994, Gregory 2001), the city as theatre or performance space has joined the list of contemporary urban metaphors. From this standpoint (urban) culture and urban space are not primarily signified by distinctive assemblages of materials or characteristic behavioural traits. Instead they exist because they are enacted and reenacted by people through various everyday practices of display – conscious and unconscious, ritualised and improvised. (Borden 2001) These practices both confirm and test the boundaries of acceptable behaviour and invest otherwise prosaic acts with a sense of enchantment. They also structure identities of the performers and the performance space (i.e. the city). The synergistic dynamism that is often associated with large cities – and much valued by their residents – may simply be a reflection of the aggregated performances that occur therein.

Such concerns seem a far cry from the material preoccupations of researchers who study climatic risks but the conceptual distance is perhaps not so great. For example, during the 1980s and early 1990s the principle task of the National Research Council's Committee on Natural Disasters was to evaluate the "performance" of sociotechnical systems during disasters (Mitchell 1987). This was an implicit recognition that vulnerability is not fixed but varies in response to many contingencies that come into play during a disaster. Likewise, recent changes in the crime rates of American cities may owe something to the advent of computer-aided GIS mapping systems that permit police to visualise – and respond to – unfolding patterns of crime instead of dispersing their resources on the basis of static (and often erroneous) assumptions about the riskiness of different neighbourhoods (Harries 1998, National Institute of Justice 2002). In this case it can be argued that an entirely new type of urban hazard management has been performed into existence by means of a new sociotechnical system.

Performance theory can be helpful to analysts of urban hazard in various ways. For example, it suggests that the identity of a city is not a given; identity is contingent, learned and malleable. Identities ought to be of concern to urban managers because they influence the interpretation of urban vulnerability. For example, in American cities a typical post-disaster enactment of identity involves public leaders vowing to remake damaged communities bigger and better. This assertion lends urgency to processes of rebuilding and material investment for their own sake and usually eclipses other important considerations such as preventing the recreation of a potential for future disasters. The process by which some identities come to be reinforced while others are ignored or rejected seems to be connected with standards and practices of public performance among leaders and public interest groups. Very little is known about these matters, although clues are sometimes glimpsed by observers; the successful contestation of Mexico City's official response to a devastating 1985 earthquake by certain groups of victims is one example. In this case victims rejected the government's claim to speak on behalf of affected populations and offered themselves as representatives of the "real" Mexico City. This suggests that there remains a vast but untapped potential for a wide range of identities to be enacted. For example, after the terrorist attacks of September 11, 2001 varied expressions of neighbourhood identity appeared throughout lower Manhattan, often signifying characteristically different orientations to urban risk and vulnerability. Most of these later disappeared when they were replaced by standardised civic enactments of identity (e.g. officially sanctioned commemorative sites) that also informed and structured subsequent public discourses about terrorism and urban vulnerability (Mitchell et al 2001).

Surprise is another concept that is important both to performance theorists and global climate change researchers. From the perspective of climate risks, surprise refers mostly to unexpected, often anxiety-producing, physical events. From the perspective of performance theory however, surprise is a valuable tool that can be employed by urban actors to introduce new ideas or new perceptions that reveal previously hidden dimensions of public issues and open new possibilities for human-induced change. In other words, surprise can be an agency as well as an outcome, a welcome antidote to stultifying orthodoxy rather than an undesired outcome. Since only a limited amount of progress has been made in formulating techniques for analysing surprise as outcome; perhaps more progress might be achieved in approaching this topic from a performance perspective.

To the author's knowledge, an evaluation of the vulnerability of a city in terms of performance criteria has not yet been attempted. Whether hazard-producing changes of urban climate might stimulate more performances, less performances or new kinds of performances or whether they would affect existing levels and forms of performance in other ways is open to debate. So too is the composition and function of the performances. Under new climate regimes would performances tend to shape a culture of risk, a culture of survival, a culture of innovation, a culture of resignation or some other normative order? At this stage it is impossible to do more than speculate about possible links between performance and vulnerability but the potential importance of these relationships might be high and is certainly worth investigating.

Cities as Muses

Etymological links between the words "city" and "civilisation" signal the fact that cities have a long history as objects of creative musing and as nurturing grounds of artistic talent. Although few contemporary global change researchers have explored literary texts, paintings or other objects d'art for ideas about urban environmental variability they can be a fertile source of useful metaphors (Nordstrom and Jackson 2001). A small sample of writings by authors of 20th century fiction about German cities are suggestive of the possibilities. These include quotations from works by Thomas Mann, Herman Hesse, Vladimir Nabokov, Wolfgang Koeppen and W. G. Sebald (Box 2.6.4). Two of these accounts highlight links between urban places and characteristic hazards while the other three focus on urban hazards as liminal events.

Mann links a particular city (Venice) to a specific hazard (plague) by skilfully using images of lassitude, decay and destruction that can be extracted both from the place's architectural palimpsest and from the fascination with material expressions of urban decline that has been a periodic characteristic of Western art. Sebald employs elegiac recollection to make a similar connection between a different kind of hazard (coastal erosion), an urban community that no longer exists (Dunwich), and – by extension – the potential for dissolution that attends all settled spaces. Similar associations of urban places and hazards have pulsed through most cultures, often informing the paradigms that are constructed by scientists and social scientists. In the aftermath of the terrorist attacks in New York City, we may well be witnessing another such episode (Mitchell 2003).

Hesse, Nabokov and Koeppen call attention to the role of urban hazards in marking boundaries and thresholds for human lives. In what is, for the author, a liberating coming-of-age story about the windstorm that unravelled the familiar comforts of his home town, Hesse underscores the complex transformative possibilities of urban disasters for good as well as ill. A Berlin thunderstorm provides the vehicle for Nabokov to parody traditional European views that associate thunderstorms with supernatural forces and simultaneously to connect readers both with a much older Greek conception of gods as flawed beings and also with the dawning realisation that humans are now taking on god-like pow-

---

Box 2.6.4 Literary views of urban natural hazards

" A luke-warm storm wind had come up .... the air was heavy and turbid and smelt of <u>decay.</u> Aschenbach .... seemed to hear rushing and flapping sounds in his ears, as though storm-spirits were abroad - unhallowed ocean <u>harpies</u> who follow those devoted to destruction, snatch away and defile their viands. For the heat took away his appetite and thus he was haunted with the idea that his food was infected."

Thomas Mann. 1911. *Death in Venice*

"Everywhere, for miles around, there were only ruins, holes, ... tree corpses with naked roots turned mournfully to the sun. A chasm had burst open between me and my childhood. This was no longer my old home...Soon afterward I left the town to become a man, to stand up against life, whose first shadows had grazed me in these days."

Hermann Hesse. 1913. *The cyclone.*

"The Thunder-god, who had fallen onto the roof, rose heavily. His sandals started slipping; he broke a dormer window with his foot, grunted, and, with a sweep of his arm grasped a chimney to steady himself. He slowly turned his frowning face as his eyes searched for something – probably the wheel that had flown off its golden axle. Then he glanced upward, his fingers clutching at his ruffled beard, shook his head crossly – this was probably not the first time that it happened – and, limping slightly, began a cautious descent."

Vladimir Nabokov. 1924. *The thunderstorm.*

"Even the storms seemed to be man-made here, an artificial entertainment in the restoration businesses of Fatherland and Sons Inc."

Wolfgang Koeppen. 1953. *The hothouse.*

"Dunwich with its towers and many thousand souls, has dissolved into water, sand and thin air. If you look out from the cliff-top across the sea towards where the town once must have been, you can sense the immense power of <u>emptiness.</u> Perhaps it is for this reason that Dunwich became a place of pilgrimage for <u>melancholy</u> poets..."

W. G. Sebald 1999. *The rings of Saturn.*

ers as agents of environmental change. The latter theme claims center stage in Koeppen's characterisation of storms over post-World War II Bonn as part of a human-engineered project for reforming the national identity of contemporary Germany.

While there are overlaps with scientific concerns and issues in all of these musings, there is no one-to-one correspondence between artistic and scientific perspectives. This should sensitise environmental change researchers to the need for carefully examining the conceptual containers into which such work is placed and the means by which the contents of those containers are made intelligible to lay publics that may hold quite different and arguably legitimate notions about them.

How – if at all – are muses vulnerable? As with the vulnerability of performance there is no easy answer to that question at this time. Perhaps the historic record of societies that were stressed by environmental changes offers some clues. It has been observed that artistic achievements and creative outputs are sometimes correlated with adversity, at last up to a point. But whether creative responses like those act as solace, or as inspiration, or as aids to action by the larger society, or in some other manner is uncertain. Doubtless, in advanced urban societies the role of the muse assumes special importance. Symbolic vulnerabilities – especially those that affect cultural treasures – often receive spectacular attention. For example, Italian governments have long placed a premium on protecting components of the national patrimony. International campaigns to counter threats to architectural or artistic muses (including museums) of Venice, Florence, Abu Simbel and Manhattan come readily to mind.

Summary and conclusions

Research on environmental change has long embraced the notion that risk-bearing change agents interact with vulnerable entities to produce outcomes that are more or less threatening or beneficial to human society. While knowledge about risks has become sophisticated, research on urban vulnerabilities is still at a relatively early stage. This paper has sought to deepen understanding of urban vulnerability by shifting attention away from the concept of vulnerability as a property of phenomena (e.g., people, buildings, ecosystems) to functional dimensions of vulnerability that have been largely overlooked or ignored. In other words, here the emphasis is on what vulnerability DOES to the functioning of an entire metropolis, rather than on what vulnerability IS for impacted individuals and groups within it. This shift has advantages for environmental change analysts because it offers a way of viewing vulnerability at an (urban) scale that is intermediate between the generalised analysis of global change impacts or consequences that is associated with top-down scientific perspectives and the highly specific measures of local vulnerability that emerge from empirical field research to inform bottom-up perspectives. Since this is the scale at which most humans live (as residents of urban communities) and the scale at which decisions about the management of environmental change tend to involve most inputs from affected publics, it is also the scale for which it is important to develop appropriately scaled concepts and methods of analysis.

The assessment of functional urban vulnerability presents particularly challenging problems because there are so many different conceptions of cities, so many potentially salient urban functions and so many processes of urbanisation that no single functional model of cities will suffice. Therefore, six different metaphor-based models of cities are used to illustrate the range of possibilities. These include conceptions of cities as machines, organisms, learning systems, regulated territories, performance spaces and muses. The models draw their inspiration respectively from contemporary research in physical science, ecological science, behavioural science, political economy, postmodernist thought and more traditional scholarship in the humanities.

Taken together, these models suggest that cities perform many functions that are differentially vulnerable to environmental change risks. Therefore the task of assessing urban impacts of environmental change cannot be viewed as a binary process that involves merging one set of information about risk with another set of information about vulnerability. Instead, there exists a family of vulnerabilities that are to some extent similar – because they all involve the potential for loss – but are also different because the meanings attached to loss and the collective functionalities of loss are not the same across the spectrum of vulnerabilities. Risks interact with different vulnerabilities in potentially different ways. This makes the task of integrating research on urban vulnerability – and by extension urban environmental change – particularly difficult.

Whether it will be possible to merge the varied vulnerabilities to produce a composite vulnerability score or other synthesis measure, is open to debate. Perhaps more important, whether such a measure would have any practical utility is seriously in question. In any case, to be comprehensive in scope and representative of the perspectives of different urban interest groups, the process of urban vulnerability assessment needs to draw upon the experience and expertise of specialists from branches of the sciences, social sciences and humanities that have not heretofore been much involved in global change research. This is a task that should be embarked on quickly because the pace of research on global environmental change is rapid and there is much ground to be made up if society is to possess a fully rounded understanding of urban vulnerability that can provide a reliable basis for policy making.

References

Armstrong MJ (2002) Urban hazard mitigation: increasing human security through re-assessing the steps towards livable communities. ICF Consulting, Fairfax. http://www.icfconsulting.com/Publications/doc_files/Urban_Hazard_Mitigation.df

Borden I (2001) Skateboarding, space and the city: architecture, the body and performative critique. Berg Publications Ltd, Oxford

Demeritt D (2002) What is the 'social construction of nature'? A typology and sympathetic critique. Progress in Human Geography 26(6):766-89

Eder K (1996) The social construction of nature: a sociology of ecological enlightenment. Sage Publications, Beverly Hills

Gandy M (2002) Concrete and clay: reworking nature in New York City. MIT Press, Cambridge

Goffman E (1959) The presentation of self in everyday life. Doubleday, Garden City

Gregory D (2001) Producing and performing space: rumours from Cairo. Alexander von Humboldt Lecture. June 6, 2001, Catholic University, Nijmegen

Harries K (1999) Mapping crime: principle and practice. National Institute of Justice research report. Washington, D.C., NCJ 178919

Inoguchi T, Newman E, Paoletto G (eds) (1999) Cities and the environment: new approaches for eco-societies. United Nations University Press, Tokyo/New York/Paris

McCarthy JJ et al. (eds) (2001) Climate Change 2001: impacts, adaptations, and vulnerability. Contributions of Working Group II to the Third Assessment Report. Cambridge University Press, Cambridge

Lefebvre H (1991) The production of space. Blackwell, Oxford

Metropolitan East Assessment Project of the U.S. National Climate Assessment Program. http://metroeast_climate.ciesin.columbia.edu/

Mitchell, JK (1987) National Academy of Sciences/National Research Council post-disaster surveys: Their applicability for mitigation purposes. Post-disaster response and mitigation of future losses. In: American Bar Washington, D.C. Association (1985) Proceedings of the international symposium on housing and urban redevelopment after natural disasters. October 23-26 1985, Bar Harbour, FL, pp 90-93

Mitchell JK (1998) Urban metabolism and disaster vulnerability in an era of global change. In Schellnhuber HJ, Wenzel V (eds) Earth system analysis: integrating science for sustainability. Springer, Berlin/Heidelberg/New York, pp 359-380

Mitchell JK (ed) (1999) Crucibles of hazard: megacities and disasters in transition. United Nations University Press, Tokyo

Mitchell JK et al. (2001) Field observations of Lower Manhattan in the aftermath of the World Trade Center disaster: September 30, 2001. Quick Response Research Project 139 (Natural Hazards Research and Applications Information Center, University of Colorado, Boulder). http://www.Colorado.EDU/hazards/qr/qr139/qr139.html and radical interpretations of disaster, Anglia University, U.K. http://www.anglia. ac.uk/geography/radix

Mitchell JK (2003) The fox and the hedgehog: myopia about homeland vulnerability in US policies on terrorism. Research in social problems and public policy XII

Mitchell JK (2003) Urban vulnerability to terrorism as hazard. In: Cutter SL, Richardson D, Wilbanks TJ (ed) The geographical dimensions of terrorism. Routledge, New York

National Institute of Justice (2002) Mapping and analysis for public safety program. http://www.ojp.usdoj.gov/nij/maps/

Nordstrom, KF, Jackson NL (2001) Using paintings for problem-solving and teaching physical geography: examples from a course in coastal management. Journal of Geography 100:141-151

Pielke, Jr. RA, Pielke, Sr. RA (1997) Hurricanes: their nature and impacts on society. John Wiley and Sons, Chichester/New York, pp 47-49

Platt RH, Rowntree RA, Muick PC (eds) (1994) The ecological city: preserving and restoring urban biodiversity. University of Massachusetts Press, Amherst

Rapport N (1998) Hard sell: commercial performance and the narration of the self In: Hughes-Freeland F (ed) Ritual, performance, media. Association of Social Anthropologists of the Commonwealth, ASA monograph 35, Routledge, London, pp 177-193

Turner V (1986) The anthropology of performance. PAJ Publications, New York

White RR (1994) Urban environmental management: environmental change and urban design. John Wiley and Sons, Chichester/New York

Wolman A (1965) The metabolism of cities. Scientific American, September:179-88

Yeatman A (1994) Postmodern revisionings of the political. Routledge, London

2.7 Energy in a Sustainable Development Perspective

Alfred Voß

Institute of Energy Economics and the Rational Use of Energy, University of Stuttgart
Heßbrühlstr. 49a, 70550 Stuttgart

The Challenges

If the historic trends of the world continue unfettered, mankind will be faced with catastrophes that are currently inconceivable. These catastrophes will be triggered off by hunger and poverty, by the destruction of the natural resource base for life or by man-made destabilisation of the earth's climate. These problems are all directly linked to the energy supply system

- since providing an increasing amount of energy services is a necessary pre-condition for eradicating hunger and poverty and even limiting the global population increase,
- since about three-quarters of anthropogenic emissions of CO_2 are released by the energy system,
- since today's energy system consumes the major share of finite fossil resources and is the single most important source of air pollution,
- since securing the economic productivity of developed countries will not be possible without a functioning energy infrastructure and competitive energy prices.

This is why energy issues featured prominently in the discussions of the earth summits from the United Nations Conference on Environment and Development in Rio de Janeiro in 1992 to the World Summit on Sustainable Development in Johannesburg in 2002. As far as these energy related challenges are concerned there is widespread agreement that they exist. However, there is a lack of consensus with respect to the targets and ways to achieve the necessary changes. Controversial, sometimes even contradictory opinions exist amongst important groups in society – at least in the industrialised countries – on the course to be taken. Resolving the `trilemma' between the economic aspirations of a rapidly expanding global population, ensuring available resources, and the environment, is one of the most critical challenges of the twenty-first century.

Access to energy is of central importance to human welfare, economic and social development including poverty alleviation. Although global primary energy use grew in the past by about 2 percent a year, on a per capita basis, the increase in total primary energy use has not resulted in any notable reductions in the difference between energy services access in industrialised countries and developing countries. Slightly more than one billion people in the industrialised countries (about 20 percent of world's population) consume nearly 60 percent of the total energy supply whereas the five billion people in developing countries consume the other 40 percent of total energy supply. Affordable commercial energy is still beyond the reach of one-third of humanity. Although energy intensity in modern economies is decreasing, more energy will clearly be needed to fuel global economic development and to deliver opportunities to the billions and still increasing number of people in developing countries who do not have access to adequate energy services.

The rate of global commercial energy consumption is a thousand times smaller than the energy flows from the sun to the earth. Primary energy use is reliant on fossil fuels (oil, natural gas and coal), which represent nearly 80 percent of total consumption. Nuclear power contributes about 7 percent, and hydro power and new renewable each contribute about 2 percent. Traditional, non commercial sources of energy, like firewood, are the dominant fuel source in low income developing countries, and account for about 10 percent of the total fuel mix.

The environmental impacts caused by energy use is not a new issue. For centuries, wood burning has contributed to deforestation. In the past 100 years, during which world's populations more than tripled and the use of fossil fuels increased more than 20-fold, human environmental impacts grew from being locally based to generating global impacts. At every level (local, regional, global), the environmental consequences of current patterns of energy use contribute a significant fraction of human impacts on the environment. Energy provision involves large volumes of material flows, and large-scale infrastructure to extract, process, store, transport and use it, and to handle the waste. Particulate matter, which is both emitted directly and formed in the air as a result of the emissions of gaseous precursors in the form of oxides of sulphur and nitrogen, and hydrocarbons impacts human health. So too does ozone, which is formed in the troposphere from interactions among hydrocarbons, nitrogen oxides, and the sunlight. Precursors of acid deposition can be transported over hundreds of kilometres and the resulting acidification is causing significant damage to natural ecosystems, crops and human made structures. Large hydropower projects often raise environmental issues related to flooding, whereas in the case of nuclear power, issues such as waste disposal raise concern. On the global scale, the possibility of significant climate change, caused primarily by greenhouse gas emissions from fossil fuel burning presents a great challenge for the future of human civilisation. There is growing evidence that much of world's energy is currently produced and consumed in ways that can not be sustained if technology development is static and if overall quantities consumed increase substantially.

Among other aspects, the risks and uncertainties of the global warming problem have led to a resurgence of interest in sustainable development. Since the United Nations Conference on Environment and Development in Rio de Janeiro in 1992 the concept of sustainable development as a model of environmentally compatible and socially acceptable development of human activities has gained widespread attention. Despite this growing interest, sustainable development is often so broadly defined, that the concept is used by different people to mean very different things. The energy debate being a prominent example. To prevent the concept of sustainability from becoming a mere buzz-word, there is a need to define what the concept of sustainable development means for the energy system in concrete terms and how it can provide guidance on the comparative assessment of energy supply options with regard to a sustainable provision of energy.

The concept of sustainable development: What does it mean for the energy system?

According to the Brundtland Commission, and the Rio Declarations, the concept of 'sustainable development' embraces two intuitively contradictory demands, namely the sparing use of natural resources and further economic development. The Brundtland Commission defines sustainable development as a "development that meets the needs of the present generation without compromising the ability of future generations to meet their own needs". Even if this definition has arisen against a background of environmental and poverty problems, it nevertheless represents an ethically motivated claim which is derived from considerations of fairness, keeping future generations in mind.

In a broad sense, sustainable development incorporates equity within and across countries as well as across generations, and integrates economic development, the conservation of the environment and the natural resource base for life, as well as social welfare. A key challenge of sustainable development policies is to address these three dimensions in a balanced way, taking into account their interactions and making trade-offs whenever needed. Energy has direct links with the three dimensions of sustainable development. But this broadly accepted understanding of sustainability is not very specific about how to assess sustainability, for example with reference to energy provision.

Any attempt to define the concept of sustainability in concrete terms can only be sound if – as far as the material energetic aspects are concerned – it takes the laws of nature into account. In this context the second law of thermodynamics which the chemist and philosopher Wilhelm Ostwald called "The law of happening" [Das Gesetz des Geschehens] acquires particular significance. The fundamental content of the second law of thermodynamics is that life and the inherent need to satisfy requirements is vitally connected with the consumption of workable energy and available material.

Thermodynamically speaking, life necessarily produces entropy by degrading workable energy and available material and requires a permanent input of these constituencies. But available energy and material only constitute a necessary but not sufficient condition for life supporting states. In addition to this, information and knowledge is required to create states serving life. Knowledge and information, which may be defined as "creative capacity" constitute a special resource. Although it is always limited, it is never consumed and can even be increased. Knowledge grows. Increasing "Creative capacity" that results in further technological development is of particular significance to sustainability because it allows for a more efficient use of natural resources and an expansion of the available resource base for generations to come.

Within the context of defining the concept of sustainability in concrete terms, the need to limit ecological burdens and climate change can certainly be substantiated. It becomes more difficult when confronted with the question of whether the use of finite energy resources is compatible with the concept of "sustainable development", because oil and natural gas and even the nuclear fuels which we consume today are not available for use by future generations. This then permits the conclusion that only the use of "renewable energy" or "renewable resources" is compatible with the concept of sustainability.

But this is not sound for two reasons. First, the use of renewable energy, e.g. of solar energy, also always goes hand in hand with a need for non-renewable resources, e.g. of non-energetic resources and materials which are also in scarce supply. Second, it would mean that non-renewable resources may not be used at all – not even by future generations. Given that due to the second law of thermodynamics the use of non-renewable resources is inevitable, the important thing within the meaning of the concept of sustainable development is to bequeath to future generations a resource base which is technically and economically usable and which allows their needs to be satisfied at a level at least commensurate with that which we enjoy today.

However the energy and raw material base available is fundamentally determined by the technology available. Deposits of energy and raw materials which exist in the earth's crust but which cannot be found or extracted in the absence of the requisite exploration and extraction techniques or which cannot be produced economically cannot make any contribution towards securing the quality of life. It is therefore the state of the technology, which turns valueless resources into available resources and plays a joint part in determining their quantity. As far as the use of limited stocks of energy is concerned this means that their use is compatible with the concept of sustainability as long as it is possible to provide future generations with an equally large energy base which is usable from a technical and economic viewpoint. Here we must note that in the past the proven reserves, i.e. energy quantities available technically and economically, have risen despite the increasing consumption of fossil fuels. Moreover, technical and scientific progress has made new energy bases technically and economically viable, for instance nuclear energy and part of the renewable energy sources.

As far as the environmental dimension of sustainability is concerned, the debate should take greater note of the fact that environmental pollution, including those connected with today's energy supply, are primarily caused by anthropogenic flows of substances, by substance dispersion i.e. the release of substances into the environment. It is not, therefore, the use of the working potential of energy which pollutes the environment but the release of substances connected with the respective energy system, for instance the sulphur dioxide or carbon dioxide released after the combustion of coal, oil and gas. This becomes clear in the case of solar energy which, with the working potential – solar radiation – it makes available, on the one hand, the principle source of all life on earth but is also, on the other hand, by far the greatest generator of entropy, because almost all of the sun's energy is radiated back into space after it has been devalued to heat at the ambient temperature. Since its energy, the radiation, is not tied to a material energy carrier, the generation of entropy does not produce any pollution in today's sense of the word. This does not, of course, exclude the release of substances and associated environmental pollution in connection with the manufacture of the solar energy plant and its equipment.

The facts addressed here are significant because they entail the possibility of uncoupling energy consumption and environmental pollution. The increasing use of workable energy and a reduction in the burdens on the climate and the environment are not, therefore, a contradiction in terms. It is the emission of substances that have to be limited, not the energy uses themselves, if we want to protect the environment.

In addition to expanding the resource base available, the economical use of energy or rather of all scarce resources is, of course, of particular significance in connection with the concept of "sustainable development". The efficient use of resources in connection with the supply of energy does not only affect energy as a resource, since the provision of energy services also requires the use of other scarce resources including, for instance, non-energetic raw materials, capital, work and the environment. Additionally efficient use of resources within the concept of sustainability also corresponds to the general economic efficiency principle. Both provide the basis for the conclusion that an energy system or an energy conversion chain for the provision of energy services is more efficient than another if fewer resources, including the resource environment, are utilised for the energy service.

In the economy costs and prices serve as the yardstick for measuring the use of scarce resources. Lower costs for the provision of the same energy service mean an economically more efficient solution which is also less demanding on resources. The argument that can be raised against using costs as a single aggregated indicator of sustainability with respect to resource usage performance is, that the external effects of environmental damage for instance are not currently incorporated in the cost figures. This circumstance can be remedied by an internalisation of external costs. Without addressing the problems associated with external cost valuation here, the concept of total social costs that combine the private costs with the external ones could serve as a suitable yardstick for measuring the utilisation of scarce resources.

Total social costs could therefore serve as an integrated indicator of the relative sustainability of the various energy and electricity supply options and it would be appropriate if, in this function, they were again to be afforded greater significance in the energy policy debate. Furthermore, cost efficiency is also the basis for a competitive energy supply, in helping economic development and adequate employment and as well as being the key to avoiding intolerable climate change. Both of these issues are central aspects of the concept of "sustainable development".

Following this clarification of the concept of sustainable development with regard to the provision of energy services an attempt is made to demonstrate, how energy supply options can be compared with regard to their relative sustainability. The assessment will be based on a set of sustainable development indicators, including emissions to the environment, the requirement of both energetic and non-energetic non-renewable resources, health impacts and total social costs as an integrated sustainability criterion.

Sustainability of energy option: A comparative assessment

The approach of Life Cycle Assessment (LCA) provides a conceptual framework for a detailed and comprehensive comparative evaluation of energy supply options with regard to their resource, health and environmental impacts as important sustainability indicators. Full scope LCA considers not only the direct emissions from power plant construction, operation and decommissioning, but also the environmental burdens and resource requirements associated with the entire lifetime of all relevant upstream and downstream processes within the energy chain. This includes exploration, extraction, fuel processing, transportation, waste treatment and storage. In addition, indirect emissions originating from material manufacturing, the provision and use of infrastructure and from energy inputs to all up- and downstream processes are covered. As modern technologies increasingly tend to reduce the direct environmental burdens of the energy conversion process, the detailed assessment of all life cycle stages of the fuel chain is a prerequisite for a consistent comparison of technologies with regard to sustainability criteria.

The LCA was carried out for a set of important electricity generation options, which is considered representative of technologies currently operated in Germany. The following figures and tables will summarise results for some of the key impact categories. They should serve as an illustrative example how to assess energy-related technologies with regard to sustainability

Cumulative energy requirements

The generation of electricity is associated with relatively intensive energy consumption for power plant construction, and – in the case of fossil and nuclear energy sources – also for fuel supply and waste treatment. The cumulative energy

Table 2.7.1 Cumulative energy requirements (CER) and energy payback periods (EPP)

	CER (without fuel) [kWh$_{Prim}$/kWh$_{el}$]	EPP [month]
Coal	0,28 - 0,30	3,2 - 3,6
Lignite	0,16 - 0,17	2,7 - 3,3
Gas CC	0,17	0,8
Nuclear	0.07 - 0,08	2,9 - 3,4
PV	0,62 - 1,24	71 - 141
Wind	0,05 - 0,15	4,6 - 13,7
Hydro	0,03 - 0,05	8,2 - 13,7

requirement as shown in Table 2.7.1 for different power generation systems includes the primary energy demand for the construction and decommissioning of the power plant as well as for the production and supply of the respective fuels. The energy content of the fuel input is not included in the figures.

The indirect primary energy input per produced kWh of electricity for hydro, wind and nuclear systems is in the range of 0.03 to 0.15 kWh. For natural gas and coal the necessary energy input per produced unit of electricity is in the range of 0.16 to 0.30 kWh which is basically determined by the energy required for the extraction, transport and processing of the fuel. The corresponding figures for today's photovoltaic systems are 0.62 to 124 kWh. This is also reflected in the energy amortisation time which is approximately 6 to 12 years in the case of photovoltaic systems using today's technology and is by far the longest compared to any of the other systems.

Raw material requirements

Electricity production involves consumption of non-energetic raw materials such as iron, copper or bauxite. Sustainability also means the efficient use of such resources. Table 2.7.2 shows the cumulated resource requirements of the power generation systems considered here for selected materials. It covers the raw material requirements for power plant construction, fuel supply, and for the supply of other raw materials. The table only includes a small part of the various raw materials required and is therefore not a complete material balance. However, results indicate that the relatively small energy density of solar radiation and of the wind leads to a comparatively high material demand. This high material intensity for wind and solar energy is an important aspect with regard to energy generation costs.

Pollutant Emissions

Figure 2.7.1 compares the cumulative emissions of selected pollutants of the power generation systems considered. It is obvious that electricity generated

Table 2.7.2 Total life cycle raw material requirements

	Iron [kg/GWh$_{el}$]	Copper [kg/GWh$_{el}$]	Bauxite [kg/GWh$_{el}$]
Coal	1.750 - 2.310	2	16 - 20
Lignite	2.100 - 2.170	7 - 8	18 - 19
Gas CC	1.207	3	28
Nuclear	420 - 490	6 - 7	27 - 30
PV	3.690 - 24.250	210 - 510	240 - 4.620
Wind	3.700 - 11.140	47 - 140	32 - 95
Hydro	1.560 - 2.680	5 - 14	4 - 11

from solid fossil fuels (hard coal and lignite) is characterised by the highest emissions of SO_2, CO_2 and NO_x per unit of electricity, while emissions from the nuclear system, hydropower and wind are comparatively low. Electricity generation from natural gas causes emissions that are significantly lower than those from coal-fired systems. Although there are no direct emissions from the electricity generation stage, the high material requirements for the production of PV panels result in cumulative CO_2 and NO_x emissions of the photovoltaic fuel chain. The result is, that emissions are close to those in the gas fuel chain and far higher as far as SO_2 and particulates are concerned.

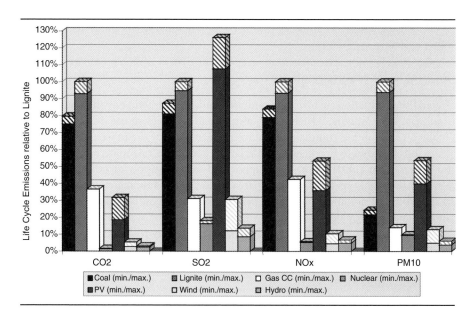

Fig. 2.7.1 Total life cycle emissions

It might be mentioned that the indirect emissions from material supply and component manufacturing are determined to a great extent by the emissions of the respective energy mix. Due to the high proportion of fossil energy in the German electricity mix, results shown in Figure 2.7.1 are not directly applicable to other countries with a different energy mix.

Human health risks

Electricity generation from fossil fuels, nuclear energy or renewable energy sources leads to an increased level of air pollution, or to an increased exposure of the population to ionising radiation, which in turn might cause an increased health risk to exposed members of the population. Using the emissions from the life cycle assessment as a starting point, health risks resulting from the operation of the energy systems considered here are assessed following a detailed impact pathway approach. For the quantification of health effects from pollutants relevant for fossil energy systems (fine particles, SO_2, Ozone) dose-effect models have been derived from recent epidemiological literature. The risk factors recommended by the International Commission on Radiological Protection (ICRP) are used to estimate effects from ionising radiation. The application of the ICRP risk factors to the very small individual dose resulting from long term and global exposure is, however, a matter of particular uncertainty and might lead to an overestimation of effects. Results of the risk assessment are summarised in the next figure. The increased death risk is presented as the loss of life expectancy in Years of Life Lost (YLL) per TWh.

Figure 2.7.2 shows that electricity generation from coal and lignite lead to the highest health risks of the power generation systems considered, while power generation from nuclear systems, wind and hydro energy is characterised by the lowest risk. Due to the high emissions from the materials, risks from photovoltaic systems are higher than the risks from natural gas-fired power plant. Results for the nuclear fuel chain include the expected value of risk from beyond design nuclear accidents, which is smaller than the current concern about major nuclear accidents in the public discussion warrants. However, the expected value of risk is not necessarily the only parameter determining the acceptability of a technology. Different evaluation schemes that take into account risk aversion or a maximum tolerable impact might lead to a different ranking of technologies.

External costs

External costs resulting from impacts on human health, agricultural crops and building materials are considered quantifiable with a reasonable level of uncertainty, but impacts on ecosystems and in particular potential impacts from global climate change can not be readily quantified, based on current knowledge. As a result an economic valuation of the potential impacts is very uncertain. In these cases, marginal abatement costs for achieving policy-based environmental targets (German CO_2-reduction targets in the case of global warming, and SO_2- and NO_x-

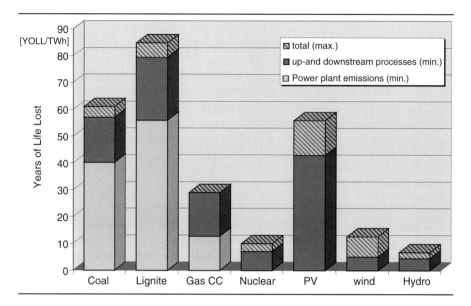

Fig. 2.7.2 Health risks of energy systems

targets derived from the European Commission's strategy to combat acidification for ecosystem protection) can be used to give a rough indication of the potential damage costs. Using the detailed Life Cycle Inventories as reference input data, the marginal external cost estimates are based on applications of the "impact pathway approach", established in the EU ExternE Project. The "impact pathway approach" models the causal relationships from the release of pollutants through their interactions with the environment to a physical measure of impact determined through damage functions and, where possible, a monetary valuation of the resulting welfare losses. Based on the concept of welfare economies, monetary valuation follows the approach of "willingness-to-pay" for improved environmental quality. The valuation of increased mortality risks from air pollution is based on the concept of 'Value of Life Year Lost'.

External costs calculated for the reference technologies are summarised in Figure 2.7.3. For the fossil electricity systems, human health effects, acidification of ecosystems, and the potential global warming impacts are the major source of external costs. Although, the power plants analysed are equipped with efficient abatement technologies, the emission of SO_2 and NO_x due to the subsequent formation of sulphate and nitrate aerosols leads to considerable health effects due to increased "chronic" mortality. A comparison between the fossil systems shows that health and environmental impacts from the natural gas combined cycle plant are much lower than from the coal and the lignite plant.

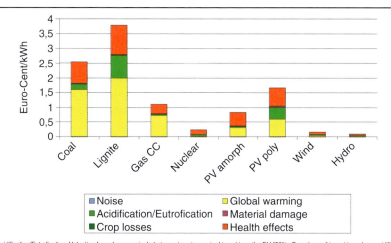

Acidification/Eutrofication: Valuation based on marginal abatement costs required to achieve the EU "50%- Gap closure" target to reduce acidification in Europe
Global warming: Valuation based on marginal CO$_2$-abatement costs required to reduce CO$_2$-emissions in Germany by 25% in 2010 (19 Euro/tCO2) 16.1.2001

Fig. 2.7.3 External costs from different electricity generation technologies operated in Germany

External costs arising from the nuclear fuel chain are significantly lower than those estimated for the fossil fuels. Most of the radiological impacts are calculated by integrating very small individual doses over 10 000 years. The application of the ICRP risk factors in this context is at least questionable, and most likely leads to an overestimation of effects. The impact resulting from emissions of 'conventional' (i.e. SO$_2$. NO$_x$, and particles) air pollutants from the nuclear fuel chain dominate the external costs. The external costs calculated from the expected value of risk from beyond design nuclear accidents are surprisingly small, given the dominance of the potential for nuclear accidents in the public debate.

External cost of photovoltaic, wind and hydropower mainly result from the use of fossil fuels for material supply and during the construction phase. External costs from current PV application in Germany are higher than those from the nuclear fuel chain and close to those from the gas fired power plant. Impacts from the full wind and hydropower life cycle are lower than those from all other systems, thus leading to the lowest external costs of all the reference technologies considered. While the uncertainties in the quantification of external costs are still relatively large, the ranking of the considered electricity options is quite robust.

Total cost of power generation

Costs in general might be considered as a helpful indicator for measuring the use of sparse resources. It is thus not surprising that a high raw material and energy intensity is reflected in high costs. The power generation costs shown in the next figure indicate that power generation from renewable energies is associated with higher costs – much higher in the case of solar energy – than those resulting from fossil-fired or nuclear power plants. However, as discussed above, the private costs alone do not fully reflect the use of scarce resources. To account for environmental externalities, external costs have to be internalised, i.e. added to the private generation costs. Figure 2.7.4 shows that the external costs resulting from the electricity generation of fossil fuels amount from 30 percent (natural gas) to about 100 percent (lignite) of the generation costs, while for the other technologies the external costs are only a small proportion of generation costs. The internalisation of external costs might lead to competitiveness of some wind and hydropower sites compared to fossil fuels, but do not affect the cost ratios between the renewable and the nuclear systems. On the other hand it is obvious, that the full internalisation of environmental externalities would improve the competitive advantage of nuclear energy to fossil electricity production.

The results of energy and raw material requirements, life cycle emissions, risks and both external and generation costs discussed so far are based on the characteristics of current technologies. It is expected that technical development will result in a further reduction in costs and in the environmental burdens of power generation. However, this applies to all the power generation technologies considered here and has to be taken into account when accessing energy futures compatibly with sustainable development goals.

In spite of the considerable progress that has been made over the last years in life cycle assessment and external cost valuation, these are still some unresolved

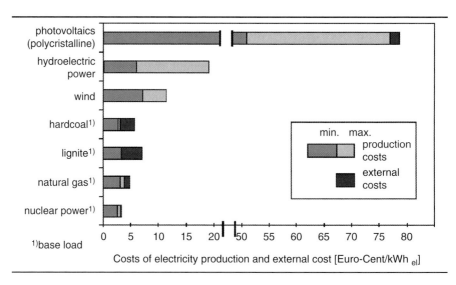

Fig. 2.7.4 Total costs of various electricity generation technologies operated in Germany

issues and partly large uncertainties. Lack of knowledge is the single most important reason for the large uncertainties related to the quantification of climate change damage costs and for the health impacts of some pollutants. The questions, of how to discount damage costs in the future, what is an appropriate discount rate as well as clarifying how to avoid risks are issues currently being debated. There is clearly a need to incorporate new scientific knowledge from epidemiological studies, global climate research and ecosystem analysis into this debate. There is also a need to take into account continuous improvements of energy technologies as well as emerging new technologies, especially when using the analysis to better inform policy making how to achieve sustainable development. Notwithstanding these caveats – and the consequent need for future work – the LCA method together with the total cost approach as an indicator for the overall resource consumption, does provide clear additional value in the decision making process with respect to a sustainable energy future.

Are sustainable energy futures possible?

Energy is central to achieving the interrelated economic, social and environmental aims of sustainable human development. But if we are to realise this important goal the development of consensus what sustainable development means in concrete terms and how to make the concept of sustainable energy provision operational is a prerequisite. We have outlined a concept of sustainable energy development with the central goal to maintain or increase the overall accessible assets (natural and man-made) available to future generations and how to assess the relative sustainability of the various energy options.

Technical progress made possible by increasing knowledge will play a key role in achieving sustainable development. Research and development provide the only systematic way to contribute to both the understanding of the earth system and the technological innovations that will be needed to meet sustainable development goals. It can extend the accessible resource base and create new categories of resources as well as increase the resource efficiency and productivity and reduce environmental impacts, thereby resulting in reduced total cost of energy service provision. Fossil, nuclear and renewable resources have considerable potential, but realising this potential will require increased technological innovation through intensified research and development, the remove of obstacles to wider diffusion and the development of economic signals to reflect environmental costs. Then the future might be much more a matter of choice than destiny. This is not to suggest that a sustainable energy provision is to be expected, only that it seems achievable. Changing the energy system in the direction of sustainability is no simple matter. It is a great challenge and a complex and long term process, one that will require concerted efforts by governments, business and members of civil society, based on a scientifically sound understanding of the concept of sustainable development.

References

European Commission (1995): ExternE – externalities of energy. Office for Official Publications of the European Communities, Luxembourg, vol 1: Summary (EUR 16520)

Friedrich R, Krewitt W (1998) Externe Kosten der Stromerzeugung. Energiewirtschaftliche Tagesfragen 48(12):789–794

Hauff V (ed) (1987) Unsere gemeinsame Zukunft: Der Brundtland-Bericht der Weltkommission für Umwelt und Entwicklung. Greven

Hirschberg S, Dones R (2000) Analytical decision support for sustainable electricity supply. In: Proceedings of VDI conference on energy and sustainable development: contributions to future energy supply. VDI, Düsseldorf, pp 168-187

Knizia K (1992) Kreativität, Energie und Entropie: Gedanken gegen den Zeitgeist. ECON Verlag, Düsseldorf

Krewitt W et al. (1998a) Health risks of energy systems. International Journal of Risk Analysis 18(4)

Krewitt W et al. (1998b) Application of the impact pathway analysis in the context of LCA – the long way from burden to impact. International Journal of Life Cycle Assessment 3(2):86-94

Marheineke T et al. (2000) Ganzheitliche Bilanzierung der Energie-und Stoffströme von Energieversorgungstechniken. IER-Forschungsbericht 74. Institut für Energiewirtschaft und rationale Energieanwendung, Stuttgart

United Nations Development Programme (2000) World energy assessment. UNDP, New York

Voß A (1997) Leitbild und Wege einer umwelt-und klimaverträglichen Energieversorgung. In: Brauch HG (ed) Energiepolitik. Springer Verlag, Berlin/Heidelberg/New York

Voß A, Greamann A (1998) Leitbild "Nachhaltige Entwicklung"-Bedeutung für die Energieversorgung. Energiewirtschaftliche Tagesfragen 48(8):486–491

Voß A (1999) Sustainability in the energy supply-relevant power generation processes on the test rig. VGB Power Tech 1:21-25

Voß A (2000) Sustainable energy provision: a comparative assessment of the various electricity supply options. In: proceedings of SFEN Conference "What Energy for Tomorrow?". Hemicycle of the council of Europe, Strasbourg, Nov 27-29. French Nuclear Society (SFEN), pp 19-27

Voß A (2002) LCA and external costs in comparative assessment of electricity chains. Decision support for sustainable electricity provision? In: Externalities and energy policies: the life cycle analysis approach. OECD, Paris, pp 163-181

Part 3

Interfaces between Nature and Society

3.1 Integrative Water Research: GLOWA Volta

Nick van de Giesen*, Thomas Berger, Maria Iskandarani,
Soojin Park, Paul Vlek

Center for Development Research, University of Bonn, Walter-Flex-Str. 3,
53113 Bonn
*Corresponding author

Introduction

Water is a theme that fits very well on the critical interface between nature and society. To fully understand the impact of changes in water supply and demand, one needs to integrate knowledge from fields as diverse as meteorology, hydrology, soil and vegetation science, medicine, economy, law, and anthropology. Our ability to deal in a scientifically sound way with all factors and feedbacks that affect the hydrological cycle is still limited. Given the extent to which our societies are dependent on sufficient, good quality water, it is quite clear that it is necessary to analyse this multi-faceted theme in a comprehensive manner. This holds especially true in the Anthropocene with its fast changes in climate, land use, trade structures, and population distribution.

Fortunately, an awareness of the necessity of integrated water science is growing rapidly as can be seen from recent global and regional initiatives, projects, and dialogues. Without attempting to be complete, we mention the Joint Water Project of the four Global Change Programs, UNESCO's HELP, the Challenge Program on Food and Water of the CGIAR, the Global Water Partnership, and the Dialogue on Water and Climate. Early on, the German Federal Ministry of Education and Research (BMBF) helped to start this trend through the GLOWA program. GLOWA, a German acronym for Global Change and the Hydrological Cycle, seeks to develop simulation tools and instruments, that allow the implementation of strategies for sustainable and future-oriented water management at the regional level (river basins of approximately 100,000km²). The objective is to predict the impact of global change, in its broadest sense, on all aspects of water availability and use. Five projects have been approved within the GLOWA program that study a total of six watersheds, aligned roughly along a North-South gradient: Elbe and Danube in Germany, Drâa and Jordan around the Mediterranean, and Ouémé and Volta in West Africa. In this chapter, we focus on experiences within the GLOWA Volta Project.

The watershed of the Volta is one of the poorest areas of Africa. Despite the presence of some precious mineral resources, average annual per capita income is estimated in the region at US $600 per year. The basin covers 400,000 km² with 42% in Ghana, 43% in Burkina Faso and the remainder in Côte d'Ivoire, Mali, Togo, and Benin Figure 3.1.1). Rainfed and irrigated agriculture is the backbone of the largely rural societies and the principle source of income. Population growth rates exceed 2.5%, placing increasing pressure on land and water resources. Improved agricultural production in the West African Savannah depends on the development of (near) surface water resources and their effective use. Such water development programs will have an impact on the availability of downstream water resources, in particular on those of the Volta Reservoir. The Volta Reservoir has the largest surface of any man-made lake in the world and provides over 95% of the electricity in Ghana. Especially the more urbanized South depends to a large extent on this energy source for its economic development. Precipitation in the region is characterised by large variability, as expressed in periodic droughts. Unpredictable rainfall is a major factor in the economic feasibility of hydraulic development schemes, as witnessed by the major 1984 drought and the power shortages that plagued Ghana in 1998. Early results from the GLOWA Volta project show a strong dependence of rainfall on the state of the land surface. Changing land use and land cover is the main global change phenomenon within the basin (van de Giesen et al. 2001).

To understand all aspects of the hydrological cycle in the Volta Basin one needs to take all these physical (atmosphere, land, water) and social aspects (population, economic development, institutions) into account. The major scientific challenge of

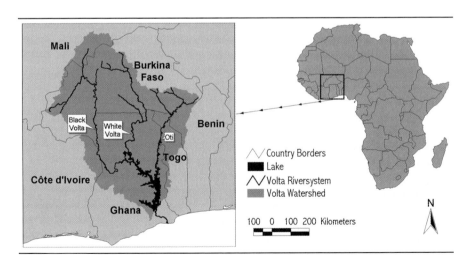

Fig. 3.1.1 Overview of the Volta Basin

the GLOWA Volta Project is, therefore, the *integrated* analysis of the physical and socio-economic factors that affect the hydrological cycle in the Volta Basin. The main content of this article concerns the experience and results of integrative research in GLOWA Volta. In the remainder of this article, we first give a general overview of the project structure, followed by a more in-depth discussion of two specific new methodologies and conclusions.

Integration in the GLOWA Volta Project

Integration of ecological, climatic, and socio-economic factors and their interactions with respect to the hydrologic cycle is the main scientific challenge of the GLOWA Volta Project. The integrated analysis will lead to a Decision Support System for the basin wide management of water resources. This DSS depends vitally on the input from many different scientific disciplines and the project is interdisciplinary in nature. The challenge for any interdisciplinary research project is on the one hand the development of a meaningful quantitative exchange of information, and on the other hand a synthesis of the different disciplinary findings, that goes beyond a mere description of the links between social, economic, agronomic, hydrological and meteorological processes. The means by which the quantitative exchange of information is pursued in this project is a set of dynamic models that capture all first order linkages between atmosphere, land surface, water, and human society.

As a first step, a panel of experts with diverse disciplinary backgrounds identified the main state variables during a workshop. According to this expert opinion, the main dynamics of water supply and demand can be captured if the following variables are modeled over space and time:

- Precipitation
- Actual evapotranspiration (ETa)
- Agricultural production
- Land use / land cover
- Population dynamics (incl. growth and urbanization)
- River flow
- Water use
- Hydro-energy
- Health
- Technological development
- Institutional development

There are, of course, many auxiliary variables that need to be measured and modelled, but the above variables represent all relevant core processes. The dynamics are addressed by fourteen sub-projects that are grouped into three research clusters: atmosphere, land use change, and water use (Figure 3.1.2). The research activities

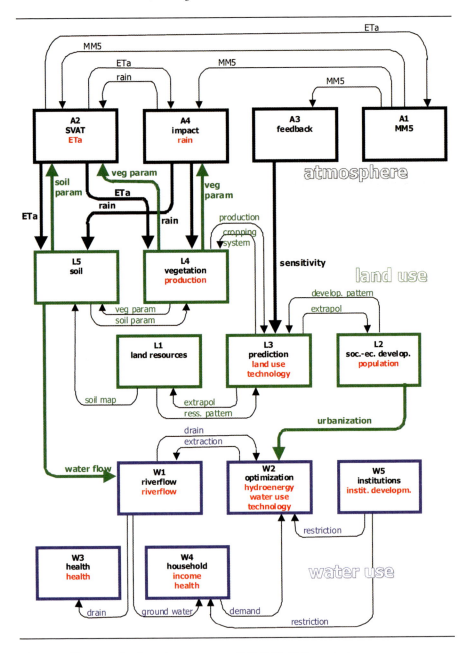

Fig. 3.1.2 Cluster and sub-project structure of GLOWA Volta project

within each cluster are undertaken by a small interdisciplinary team. By concentrating the information exchange between disciplines in small groups, scientific problems between these disciplines were solved from the on-set. Figure 3.1.2 shows how the different sub-projects are connected and also where the different key variables of the DSS will be produced.

In most instances, communication between connected sub-projects and associated models takes place between "kindred" scientists; the crop scientist exchanges information with the soil scientist who contacts the hydrologist, etc. In general, the applied models and methodologies are close enough as not to cause fundamental communication problems. Looking at Figure 3.1.2, however, one sees three foci where communication is needed between less kindred disciplines, namely the exchange between land surface and atmosphere, land use change prediction, and water use optimisation. In the upper left corner, a dense exchange of information between land surface and atmosphere is depicted. Although meteorologists and hydrologists are kindred scientists, there is a big scale gap between these two sciences that needs to be bridged to properly capture feed-back mechanisms between land use (change) and atmospheric circulation. Because this only involves natural scientists, we do not elaborate on the integration involved here. We do describe, however, the activities associated with the remaining two integrative foci, land use change and water use optimisation. Many disciplines contribute towards these two central sub-projects. In the first phase of the GLOWA Volta project (2000-2003), scientists concentrated mainly on conceptual integration or, in other words, the development of new methods of data gathering and modelling. The idea was that by converging on the conceptual integration in these focal points, it would be easier to define precise goals and to make progress with small interdisciplinary teams.

In summary, the general approach was to build in integration from the beginning of the project by designing a network of sub-projects and information exchange. Most of these sub-projects can exchange data and models with their neighbours because they are conceptually close. The more demanding conceptual integration was concentrated in three focal points. In the next two sections, we report on two of these points in which natural and social scientists developed new integrative methods of data collection and analysis. In case of land use, we present the development of a so called Common Sampling Frame that was used to collect data at village and household level related with socio-economic factors as well as soil and water quality. After the Common Sampling Frame, the development of a water use optimisation model, to which economists, hydrologists, and institutional analysts contributed, will be presented. We conclude with some remarks concerning the process of scientific integration.

Common Sampling Frame

An integrated DSS demands a coherent database, but building such a database poses methodological problems. For socio-economists, the main unit of observation is the individual household, whereas political scientists focus on decision-making processes at higher levels of social organization. Hydrologists and soil scientists use research units that are related to landscape boundaries rather than social entities.

In order to link data across different observational units and to guide surveys or field measurements, a general framework of analysis is needed. In GLOWA Volta, we developed a 'Common Sampling Frame' (CSF) based on a statistical representation of the study area using the Ghana Living Standards Survey and publicly available geospatial information. The indicators generated from the statistical procedure were used as a "map" of the socio-economic landscape, not unlike the way in which satellite data provide a map of the physical landscape.

Research questions and observation units

In GLOWA Volta, six interdisciplinary research questions were addressed with the CSF:

- How safe is access of households to water?
- What are the determinants of household water demand and consumption?
- What are household expenditures on water?
- Which water-related health problems occur?
- Is water availability linked with economic activities and migration?
- What changes in the use of land have taken place?

How do economic activities affect the spatial and temporal distribution of soil and water properties?

Evidently, answering these questions requires information from the natural sciences such as biophysical measurements and spatial data from remote-sensing techniques. In turn, the socio-economic data will inform the natural scientists' research directly ("ground-truthing" of remote sensing images) or feed into the joint integrated modelling (modelling of land use change, inter-sectoral water allocation).

For the economic sub-projects, the main unit of observation is the individual household that takes decisions regarding the use of water and land resources. The research focuses on understanding the households' choices among different feasible alternatives of action and in particular on their strategies for coping with water variability, climate and land cover changes. The institutional analysts, on the other hand, investigate decision-making processes at higher levels of social organisation, for example on community, regional and national level. The unit of observation is accordingly not the single household, but the village assembly, the water user association, etc.

The hydrologists, on the other hand, undertake their research in small, experimental watersheds (1-10km²). Soil scientists and geographers focus on landscape units that are grouped into different land use and land cover classes. In GLOWA

Volta, the size of the main units of observation of the physical scientists corresponds roughly to that of the community and district level. A perfect hierarchy of observation units of all sub-projects, however, cannot be constructed because administrative boundaries, watersheds, and land cover polygons do not fully coincide. The socio-economic and bio-physical sub-projects, including those in the atmosphere cluster, therefore agreed on an appropriate spatial resolution for exchanging data among their models. In the course of the GLOWA Volta project, a geographical information system will organise the data of all sub-projects in 9×9 km grid cells.

Advantages and disadvantages of common sampling frames

Employing a CSF provides several advantages for interdisciplinary research teams who plan to collect large amounts especially of primary data:

- The CSF can make use of a priori information for stratification and therefore tends to increase precision and reliability as compared to a pure random sampling. Especially in the case of GLOWA Volta, considerable amounts of socio-economic data with high quality and large spatial extent are available. The a priori information also helps developing research hypotheses and may guide the design of questionnaires or measurements.
- The CSF yields certain "agglomeration" benefits particularly for the project logistics. Collecting information from hierarchically linked observation units usually implies a spatial concentration of field activities. As a consequence, transport and lodging costs for enumerators and technicians can be reduced, as well as training and interpreting costs since this concentration allows building larger teams of field assistants with knowledge of local languages. Additionally, costs of transforming and exchanging data between different scientific disciplines are usually lower when collected from the same or nearby observation units.
- A hierarchical sampling frame permits the extrapolation ("grossing-up") of sample measurements to national and regional levels. The research findings at different locations can be generalised and may then be used for deriving conclusions at national or basin level. This is potentially a large advantage over a non-random and purposeful selection of observations units that allows only for comparisons of case studies or "anecdotal" evidence.

A common sampling frame implies on the other hand the following disadvantages:

- Typically, interdisciplinary teams have to invest a lot of time to agree on a hierarchical structure of observation units and operational selection criteria for stratification. The discussion process is therefore costly. In spite of this disadvantage, the discussion process might help clarifying the different viewpoints of the disciplines involved and in the longer run lead to a shared terminology or even methodology. In the case of GLOWA Volta, the construction of the CSF was the first interdisciplinary research activity where different sub-projects of the land and water cluster started working together scientifically and thereby laid the basis for future integrative research.

- The common sampling might also be inapplicable for some disciplines that will then opt not to participate. This is in particular relevant when certain sub-projects can undertake only very few and long-term measurements. A stratified random sampling implies for them a high risk of missing the observation units of their specific interest. Nevertheless, the CSF might also in these cases provide the chance of at least comparing the findings of these out-of-frame sub-projects with other sub-projects as long as information on the joint selection criteria will be collected.

Sampling frame and selection procedure

Technically, the construction of the CSF consists of the following steps (see details in Berger, et al. 2002):

- all sub-projects identify their main observation units;
- the interdisciplinary team then establishes a hierarchical structure of observation units;
- the team agrees on a sampling frame and selection criteria that reflect the interests of all sub-projects involved;
- *a priori* information is employed to stratify the universe of observation units;
- for each strata the team selects the observation and sub-observation units randomly, if possible, and calculates weighting factors;
- the sub-projects gross up their results to the study region.

As the sampling frame for the Ghanaian part of the Volta basin, the research team agreed on a merged data set taken from the Ghana Living Standards Survey (GLSS) and publicly available GIS maps. The GLSS, which was conducted by Ghana's Statistical Service with assistance from the World Bank and the European Union, provides data on various aspects of households' economic and social activities as well as community characteristics in Ghana. The survey was carried out on a probability sample of 6,000 households in 300 enumeration areas that were drawn from the Ghana population census of 1984. Of these 6,000 households sampled in the GLSS 2,240 fall within the Volta basin. The remainder of this section explains the statistical procedure in selecting the GLOWA Volta observation units on community level; full documentation can be found in Berger et al. (2002).

Compilation of a priori data base and identification of observation units

The GLSS data was first aggregated at community level and then merged with GIS information in order to gain first insights into the socio-economic and agro-ecological conditions in the Ghanaian part of the basin and their geographical distribution. This merged data set forms the basis for the statistical selection procedure. Additionally, the project scientists developed a hierarchy of observation units involving regions, districts, landscape units at community level, villages, village water sources, households and plots.

Identification of selection criteria for stratification

Within the interdisciplinary research team we decided on selection criteria that, from a theoretical point of view, potentially capture all important research aspects to the disciplines involved. Based on the preliminary analysis of the merged data set, 22 selection criteria at community level were identified that belong to the following nine categories:

- Household water security
- Agro-ecological conditions
- Agricultural intensity
- Fishing intensity
- Market access
- Household welfare
- Waterborne Diseases
- Social capital
- Migration potential

Multivariate data analysis and selection of representative communities

Table 3.1.1

Discipline	Research activity and observation unit	Purpose
Geography	Land cover recording chart of community landscape	Ground-truthing of remote-sensing images
Political sciences, Anthropology	In-depth interviews with village elders; household interviews	Institutional analysis
Economics	Household interviews	Household water demand; Migration behaviour
Agricultural Economics	Household interviews	Water and land use decisions
Soil sciences	Plot survey	Soil quality analysis
Hydrology	Bacteriological analysis of water	Water quality analysis

Initial correlation analysis of the merged data set revealed high interdependence among the selection criteria, so we used Principal Component Analysis (PCA) to derive a relatively small number of linear combinations of the original variables that retain as much information in the original variables as possible. The PCA detected high correlations among the original variables and revealed 8 components that explain about 70% of the total variance in the data. The new variables were used for the cluster analysis that identified 10 clusters (or strata). The clustering procedures gave Euclidean distances from the centroid of each cluster to their respective communities. The communities closest to the cluster centroid were then selected as representative communities proportional to their sizes.

To ensure an overlap with other GLOWA Volta sub-projects researching at locations that are not contained in the original GLSS sampling frame, additional sites were added to the sample. As a result of the sampling procedure, a list of 20 survey communities was compiled. The different disciplinary sub-projects then randomly selected their sub-observation units at lower levels, for example at household, water source or plot level.

Multi-topic survey design

A multi-topic community and household survey was designed in order to capture the socio-economic as well as natural science aspects (see Iskandarani et al. 2002). Data collected concerning the socio-economic environment contained information on:

- Institutional framework of water and land management
- Household preferences and water demand behaviour
- Agricultural production, investment behaviour and land use
- Migration decision making.

To capture human behaviour in its biophysical environment three other groups of parameters were measured in addition:

- *Land cover Classification.* A land cover recording chart of community landscape was documented and serves as ground truth for the parallel remote sensing measurements.
- *Soil Quality.* Soil samples were taken from the fields of all interviewed households and the soil quality was measured (texture, bulk density, organic content). The results help to characterize soils in the basin as well as to evaluate investment and production decisions of the households interviewed.
- *Drinking Water Quality.* E-coliform contamination at the village drinking water source and at the drinking water storage facility within the household was measured during two seasons. The results complement the socio-economic analysis of household water security in the basin.

In summary, integrating the primary data collection of natural science and socio-economic research aspects based on a common sampling frame allows the analysis of the resource users' behaviour within their individual environmental contexts. This enables integrative research to better analyse the feedback effects of household decision making on welfare levels and condition of natural resources. A common sampling frame thereby improves the understanding of the complexity of integrated economic and ecological processes, e.g. land use change and determinants of household water security. Furthermore, applying an integrated sampling frame is expected to give higher accuracy and reliability of estimations.

But there are also some limitations that have to be considered. A multi-topic survey tends to be lengthy and researchers involved in the development of the questionnaire have to be very focused in order not to overstrain the interviewers. A lengthy questionnaire may raise the risk of gaining data of lower quality, due to the fatigue of the interviewees. The design of the common sampling frame and the implementation of the survey need therefore careful preparation through intensive interdisciplinary discussion on data needs and clear thoughts about the scope for integration.

Water Use Optimisation

Overview

The second integrative focal point that we discuss in some detail is the water use optimisation model. The final product of the project is a DSS built around an optimisation routine that finds the highest returns on the scarce water resources. The "return" on water should not be seen too narrowly as immediate financial returns, but rather as a value to be determined by the users of the DSS and could also include returns that are more difficult to quantify such as social equity or ecological value. Avoiding, mitigating and – hopefully – resolving conflicts over the water resources in the Volta basin requires first of all information and insights about the underlying bio-physical and socio-economic processes. Science-based information systems that take account of these processes could then assist in quantifying the trade-offs, assess alternative allocation mechanisms, and inform policy development and analysis on interrelated water-use options.

In the case of the Volta Basin, competition over water resources takes place within and between three main sectors: irrigation, hydropower, and households. Households in the rural areas of Ghana and Burkina Faso do not consume amounts of water that substantially affect the water balance, but their social value is very high. Hydropower and irrigation, however, stand in direct competition. In typical watershed configurations, hydropower is generated in the upper reaches of the river network. The water that generates hydropower can in such watersheds still be used for irrigation in the lower lying plains. In the case of the Volta, however, irrigation development takes place on the plateaus of Northern Ghana and Central Burkina Faso, whereas hydropower is generated at Akosombo, less than 100 km from the sea (van de Giesen et al. 2001). Irrigation development in the region is a relatively diffuse process consisting of many small scale, village-level schemes. The number of such schemes has increased considerably over the past decade. Logically, one would like to know what the impact of irrigation on downstream hydropower generation would be.

In first instance, we wanted to know if it was possible to develop an integrated water use optimisation model and what problems needed to be resolved. For this purpose, a first order model was developed that was able to address the following questions:

- What is the optimal allocation of water and land resources taking into account existing physical, technical, and financial constraints?
- What are the differences between actual and return-maximizing allocation and where can additional benefits be achieved most efficiently?
- Which constraints on land- and water-resources development are most binding?

At a later stage in the project, a full coupling with land use and atmospheric models will be put in place. Once full coupling has been established, it should be possible also to address the following questions:

- How large are tradeoffs between competing water and land uses?
- What do development paths under different environmental and technological scenarios look like?
- What are feasible water and land management policies and what are their costs and likely effects?

Optimisation for the efficient allocation of resources is typically an activity for economists. In this case, however, the optimisation has to take place under hydrological and institutional constraints. Clearly, the amount of available water is limited and assessing the water availability would be the task of the hydrologist in the team. In the context of developing countries, institutional constraints are very important as well because governments are poor and do not have the means to create the institutions needed to ensure optimal water allocation.

Design

Economic optimisation models face difficulties in representing the complex spatial arrangements. Goods can normally be transported at minor costs to the place where their return is highest but in complex water systems there are additional constraints that have to be captured. In principle, one could also pump water around, but this is only economically feasible for when returns are very high such as for industrial and household water supplies. Pumping water can be excluded *a priori* on thermodynamic grounds for hydropower generation. For irrigation purposes, water may be pumped up but in Africa this does not involve moving water over large distances (>10 km). The hydrological constraints under which water allocation needs to be optimised have a clear hierarchical structure: water used upstream is not available downstream and water that is available downstream can not be used upstream.

The river network was represented in the optimisation model as a node-link network, in which nodes correspond to physical entities and the links correspond to the river stretches between these entities. The nodes included in the network were source as well as demand nodes. Each demand node is a location where water is diverted to different sites for beneficial use. Inflows into these nodes include water flows from the headwaters of the river basin and drainage between nodes. The spatial relationships between different water uses are captured in the node-link network. Water flow is routed through the model system and hydrologic balances are calculated for each node in the network.

Thirteen agricultural, domestic and industrial demand nodes were spatially connected to the basin network (Figure 3.1.3). Hydropower and agricultural demand sites are delineated according to the hydropower and irrigation facilities

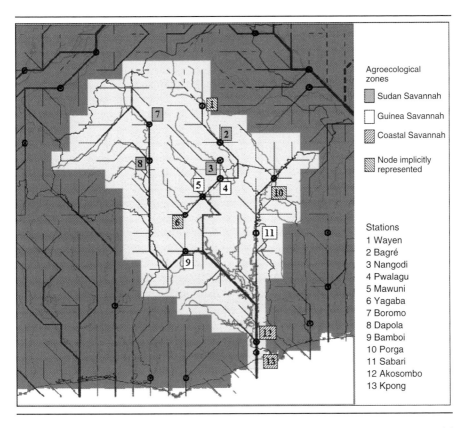

Fig. 3.1.3 Node link network for the Volta River as used in the water use optimisation model

in place. Five nodes are in the Sudan savannah, six in the Guinea savannah, and two in the Coastal savannah. The hydrological input which has been used up until the current point is ten-year average river flow. Secondary data were employed to calculate crop-water demands and irrigation efficiencies in these agro-ecological zones. For each node, a hydrologic water balance was established to inform the subsequent water optimisation model. Currently, Burkina Faso uses most of its Volta water resources for irrigated agriculture and a small percentage for municipal and industrial use. The remainder goes downstream to Ghana where the bulk of the surface water is used for hydropower generation. Domestic and agricultural use accounts for only less then one percent. Critical storage levels in the Volta Lake and frequent power shortages in recent dry years have raised concern in Ghana about plans for further irrigation development upstream in Burkina Faso.

The research questions underlying model development – analysing transboundary and inter-sector water allocation at the basin scale – are reflected in the level of abstraction and complexity of the economic model component. The supply and demand of water is represented at the aggregated level of country or region. This aggregated representation clearly abstracts away from agent heterogeneity or coordination failures. The economic agents explicitly represented in the model are the relevant water using sectors in Burkina Faso and Ghana. The model computes the optimal allocation of water across these sectors on the basis of the economic value of water. The economic value of water is subject to a series of physical, system-control, and policy constraints. The model employs a Samuelsonian objective function that corresponds to the sum of producer and consumer surplus. We assumed competitive domestic markets, but considered only own price demand elasticities for this first modelling attempt. Though the objective function is quadratic, we have implemented it in linearised form (Hazell and Norton 1986: 164). It is planned to convert this model into a general equilibrium model at a later stage of research.

Institutional constraints were built into the integrated water-optimisation model as water quota allocated to different water sectors. One example of an institutional constraint is a maximum amount of water for irrigated agriculture due to informal rules or funding limitations. The institutional analysis at local, regional as well as national level made quantification of these constraints possible. In the first simulation runs, the quantities for domestic water and energy demands as well as energy imports and exports were fixed. Quantities and prices of agricultural products, on the other hand, are endogenous assuming partial equilibria on regional markets. The institutional analysis also resulted in alternative water management scenarios. Policy scenarios that form the basis for the model runs were derived from government development plans for the next two decades. These scenarios were subsequently run by the economic-hydrologic optimisation model.

Results

The initial model results suggest that further irrigation development – an increase of irrigated area by thirty percent – would have a small effect on the reservoir level compared to rainfall and runoff variability in the Volta basin. Average energy demands in Ghana can only be met by overexploiting the water resources of the Volta Lake in the order of ten percent in terms of normal long-term storage levels. If all other factors remain unchanged, hydropower generation would have to be reduced with an amount in the order of the current Ghanaian household demand, if the Volta Lake is to be operated in a sustainable manner. Currently, a series of simulations is being run to analyse the effects of different scenarios of electricity trade and capacity expansion in Ghana and the neighbouring countries. Based on model data for the West Africa Power Pool (WAPP), developed in an USAID funded research project at Purdue University, different investment scenarios such as new hydropower and combined cycle generation in Ghana are being tested.

Though aggregation of agents facilitates practical model development, it might overlook the potentially large variation in local conditions, including climate and water-use patterns. Aggregation may, therefore, result in more favourable model outcomes than those actually existing in many local areas within the basin. As outlined in Berger and Ringler (2002), further disaggregation of the regional and country-level structure to the local/sub-regional level within a multi-scale, multi-agent framework will result in a more realistic local and sub-regional water supply and demand situation.

Constructing and running an integrated water optimisation model from the onset of GLOWA Volta has yielded several advantages. An early model based on available secondary data helped to define precise data needs in terms of water supply and demand, and thus guide the survey activities and field measurements. The model implements existing integrative knowledge of different disciplines and can gradually be enriched by updated information and new primary data collected throughout the project. Unexpected model results may also lead to new research questions and hypotheses to be addressed in the course of the project.

Using a computer model as an immediate tool for integrating, updating and communicating disciplinary knowledge is only possible thanks to recent advances in information technology. Computer memory and processing constraints are much less limiting compared to what they were twenty years ago, efficient mathematical solvers facilitate faster model development, and secondary data sets have become widely available through the Internet. These technological advances allow also a next innovative step in the GLOWA Volta project; optimisation of water use can now be carried out at different levels of aggregation and for multiple economic agents. The optimisation model has active links to a large set of sub-models that contain auxiliary information such as institutional development scenarios and crop-water demand calculations.

Conclusions and outlook

In this chapter, we presented the general integration approach in the GLOWA Volta project together with two in-depth cases in which natural and social scientists developed new methodologies. Rather than summarize, we would like to conclude this chapter with additional remarks concerning the processes that led to these results. The article ends with the outlook on future integrated research.

The overall integrative experience in the GLOWA Volta project has been positive. We set out to produce integrative insights from the onset of the project and have subsequently made interesting conceptual progress. A certain set of conditions can be found that are likely to have helped the specific implementation of the general integrative approach. First, at a practical level, the Center for Development Research, where most of the presented research took place, is an interdisciplinary institute where ecologists, economists, and anthropologists work under one roof. The simple physical proximity of the scientists involved proved to be extremely useful for the exchange of ideas and implementing integration through a trial-and error approach. Probably more important is the fact that an interdisciplinary centre tends to attract people oriented to working with scientists in other disciplines. In this case, the team consisted mainly of recently graduated post-doctoral scientists who brought along the necessary mental flexibility to try new approaches and were willing to take the associated risks.

At the conceptual level, two factors were rather important. First, in both cases, the scientists involved shared a common goal that was clear and tangible, respectively a common sampling frame and a water use optimisation model. This differs from many multi-disciplinary projects in which the individual scientists study the same object, starting out with a vague goal such as "understanding" the system at hand. Typically, in such cases the scientists first scatter to measure and model disciplinary aspects. At the end of the project, these different threads then have to be, somehow, tied together. In the GLOWA Volta case, the shared goals focused the different efforts and forced integrated thinking.

The second useful conceptual factor was the insistence on quantification and modelling. It is easy to get lost in fundamental discussions that highlight differences in underlying philosophies. This book chapter focuses on bridging gaps between natural and social sciences, but in our case the differences between economists and anthropologists were larger than those between, say, economists and hydrologists. Both economists and hydrologists are at home with mathematical problem formulations and the main technique involved (constrained optimisation) is well-known by both. The anthropologists involved in institutional analysis, however, were not comfortable with the idea of quantifying human behaviour, let alone with predicting human actions by solving equations. It took quite some group effort and, frankly, friction before a modus operandi was found. It was difficult for the anthropologists to appreciate how important obvious and simple observations really can be in constraining model solutions. Without institutional constraints, for example, the optimisation model would predict unrealistically large expansion of the irrigation sector at the cost of hydropower generation. The political reality in Ghana shows a very strong hydropower sector and a weak irrigation sector. This fact was so obvious that it did not really count as a scientific observation, yet it was a very important ingredient of the model. A certain willingness to simplify and understate the importance of disciplinary findings seems often an essential first step in integrative research. Building a (quantitative) model as a team can be helpful in putting one's own input in proper perspective.

From the above it may have already become clear that integrative science comes with a relatively large overhead. One has to agree on goals and approaches, understand the basic principles of other disciplines, and explain one's own discipline to non-experts. The development of the Common Sampling Frame brought these requirements to the fore. The soil scientist, for example, had to develop special GIS tools to aid the sampling stratification and develop specific research protocols for the physical observations made during the survey. The amount of information on soils in the Volta Basin that he received in return could, undoubtedly, have been collected by him with much less effort as an individual research activity. It was essential, however, that the social and physical information can be coupled directly in order to model land and water use dynamics.

The reward structure of present-day academia is biased towards mono-disciplinary research. The question becomes whether the extra overhead associated with interdisciplinary research is worth it. Coarsely speaking, there are three measures of success for university scientists: (1) number and quality of students, (2) number and quality of publications, and (3) research funds raised. The attractiveness of interdisciplinary research towards students varies and not much can be said about advantages and disadvantages in this respect. Very problematic is the publication aspect. Looking at, for example, the water use optimisation model, both the hydrologist and the anthropologist only contributed minimally and perhaps only the economist could publish the results in a peer reviewed journal. The number of interdisciplinary journals increases but their impact factors are low and publications in these journals are not valued as highly as publications in better known disciplinary journals. One look at an issue of Nature or Science makes it clear that successful young scientists would be smart to dig deep into special aspects of their own discipline. The best hope for interdisciplinary research lies in the rather mundane final aspect, namely fund raising. Research sponsors are more and more interested in solving real world problems. Solving such problems demands integrated research because sustainable development of our social and natural environments depends on highly complex interactions that can not be understood from a mono-disciplinary point of view. The GLOWA Program is a good example of integrated research that is promoted by a sponsor who insists on results that provide decision makers with the practical help needed to deal with all aspects of changing water demand and availability. When such programs become a standard way to provide substantial long-term research support, the research community will re-orient itself once young scientists are able to build careers as members of inter-disciplinary teams.

References

Berger T, Ringler C (2002) Trade-offs, efficiency gains and technical change – mod-elling water management and land-use within a multiple-agent framework. Quarterly Journal of International Agriculture 41(1/2):119-144

Berger T, Asante FA, Osei-Akoto I (2002) GLOWA-Volta common sampling frame: selection of survey sites. ZEF documentation of research 1/2002. http://www.glowa-volta.de

Hazell PBR, Norton, RD (1986) Mathematical programming for economic analy-sis in agriculture. MacMillan, New York

Iskandarani M et al. (2002) GLOWA-Volta household survey: documentation of questionnaire. ZEF documentation of research 2/2002. http://www.glowa-volta.de

van de Giesen N et al. (2001) Competition for water resources of the Volta basin. IAHS Publication 268:199-205

3.2 Global Carbon Cycle

3.2.1 Carbon Cycle Research as a Challenge of the Anthropocene

Karin Lochte

Institute for Marine Research at the University of Kiel, Düsternbrooker Weg 20, 24105 Kiel

Introduction

The carbon cycle has a very distinctive role in the Earth system due to the fact that organic carbon components represent the basic form of energy storage by biological processes and that they form the building blocks of all life on Earth. Carbon is also involved in many physicochemical processes on all time scales. Biological processes have caused the accumulation of ancient organic carbon deposits in the form of oil, coal, peat, methane gas or gas hydrates over the millennia as well as massive deposits of inorganic carbon (e.g. limestones). The recent reactivation of these fossil fuel deposits for energy production by man over the very short time span of less than a century has led to the present sharp rise in the concentration of carbon dioxide (CO_2) in the atmosphere as shown in Figure 3.2.1.1.

The concentration of CO_2 in the atmosphere is one component, amongst several, influencing the heat balance of Earth over long time scales. Changes in its concentration are generally associated with changes in Earth's climate. Therefore, much attention in recent years has been focussed on understanding variations in CO_2 concentration in the atmosphere. It has to be borne in mind, however, that other "greenhouse" gases are concomitantly released into the atmosphere enhancing the effect of CO_2 (Prather et al. 2001). The most important ones are water vapour, methane (CH_4), nitrous oxide (N_2O) and tropospheric ozone (O_3). They either originate directly from human activities, such as burning of plant biomass or release of methane from rice paddies and cattle, or their production is caused by changing environmental conditions, such as changes in the water cycle or emission of methane and dinitrogenoxide from suboxic areas on land and in the sea. The effect of these gases on the heat balance of Earth is amplified by changes in water vapour content in the atmosphere. Furthermore, aerosols may add or diminish the effects of "greenhouse" gases (Penner et al. 2001). Depending on the type of aerosol and its distribution, it can either reflect solar radiation back

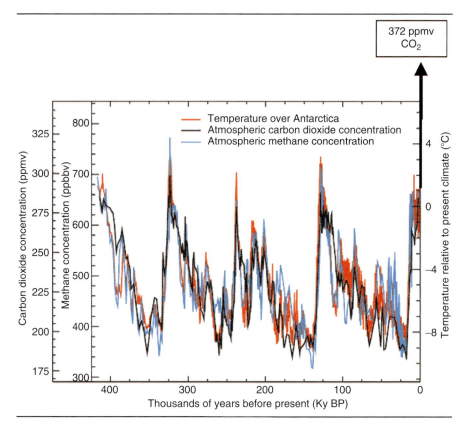

Fig. 3.2.1.1 Sharp rise of CO_2 concentration in the atmosphere

into space providing a cooling effect or it can absorb solar radiation and thus heat up certain strata of the atmosphere. Backscattering of solar radiation by clouds (cloud albedo) is a major feedback in the atmosphere to regulate the surface temperature of the Earth. At present, the heating effect of the "greenhouse" gases is thought to be compensated by the backscattering of solar radiation by clouds and aerosol particles. Recent estimates indicate that the heating impact of "greenhouse" gases may be reduced by 65%-80% due to albedo effects (Crutzen and Ramanathan, Dahlem Conference May 2003, in press). These authors argue that it is not likely that aerosols will increase at the same rate as CO_2 concentrations in the atmosphere and, hence, the heating effect may rise more rapidly in future.

   High CO_2 concentrations in the atmosphere, however, have also quite different effects on the environment. They increase photosynthesis in land plants, reduce the pH of water and can alter calcification processes. Experiments have shown that carbonate formation by marine organisms, including plankton, algae and corals, is inhibited at concentrations of ca. 800 ppm expected for the year

2100 (Kleypas et al. 1999, Riebesell et al. 2000). In fact, models predict a drastic reduction in carbonate formation (Figure 3.2.1.2) with unknown consequences for the marine ecosystem. It would endanger coral reefs, decrease their high bio-diversity and may deteriorate the stabilisation by the reefs for coasts and tropical islands. It would also change the composition of plankton organisms in the open ocean with possible impacts on the production of fish.

In addition to the above mentioned direct influences of rising CO_2, indirect effects are suspected from climatic changes triggered by the increasing atmospheric concentrations. At present, rising global temperatures are being observed that are the result of the combined effects of "greenhouse" gases, water vapour, aerosols and clouds. These components are linked by various feedback processes, which enhance or compensate each other. Therefore, the understanding and modelling of such complex interacting systems is a major challenge for future research which requires to consider the linkages within the Earth system. Global average temperature increase predicted by different models range from 1.7 to 4.2°C with a mean of 2.8°C over the next 100 years. Sea level rises due to thermal expansion of sea water and melting of glaciers are predicted to amount to 0.48m (range 0.09-0.88m) over the same time span as a result of global warming. Such climatic changes will alter hydrological cycles, ocean stratification and current systems.

The present knowledge of carbon flows and their effects on our climate is compiled most comprehensively in the IPCC (Intergovernmental Panel on Climate Change) reports (IPPC 2001). After a decade of basic research into the pivotal role of the element carbon in the Earth system, the international scientific community has now launched into a new effort to link the present understanding of the natural carbon cycle with the behaviour of human societies. This "Global Carbon Project" (GCP, see http://www.globalcarbonproject.org/) has defined the following central issues for the next decade of research:

1. Patterns and variability: What are the current geographical and temporal distributions of the major stores and fluxes in the global carbon cycle?
2. Processes and interactions: What are the control and feedback mechanisms – both anthropogenic and non-anthropogenic – that determine the dynamics of the carbon cycle?
3. Management of the carbon cycle: What are the likely dynamics of the carbon-climate system into the future and what points of intervention and windows of opportunity exist for human societies to manage this system?

The GCP is a joint project of the four global change programmes IGBP, WCRP, IHDP and DIVERSITAS. Its main function is to address the above questions by integration of knowledge and data obtained by other research programmes. It has the task to link natural and human sciences with the aim to better understand the carbon cycle and to assess mitigation and adaptation options for society.

This article will firstly give a brief overview over the present understanding of the variability and of critical processes in respect to atmospheric carbon concentrations. In the following some of the research issues will be highlighted that were

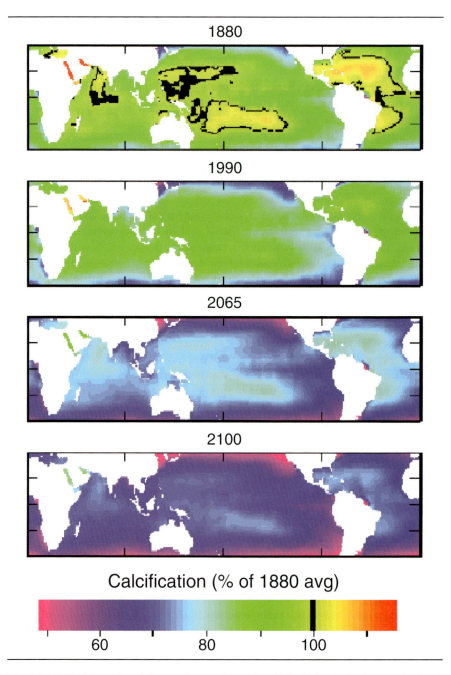

Fig. 3.2.1.2 Model results of future changes in marine biological calcification on the basis of predicted CO_2 concentrations and changes in sea water pH in the year 2100. The figure shows the percentage of calcification in 2100 relative to the reference year 1880 (figure from Kleypas et al. 2000).

identified in several workshops as being of particular relevance to link natural and socio-economic sciences in respect to carbon research.

Variability of atmospheric CO_2 concentrations in the past

During earth's history the concentrations of CO_2 in the atmosphere have changed strongly. Very high concentrations (of > 3000 ppm) were found between 600 to 400 Myr BP (Cambrium, Ordovicium, Siluran) and high concentrations also occurred between 200 to 150 Myr BP (Triassic, Jurassic) when drastically different climatic conditions prevailed (Berner 1997, Hayes et al. 1999). In the past 60 Myr (Tertiary) the concentrations gradually decreased and remained below 300ppm for the last 23 Myr (Pearson and Palmer 2000). During the last 420 kyr they ranged between 190 ppm (glacials) and ca. 280 ppm (interglacials) (Figure 3.2.1.1) (Petit et al. 1999). The fluctuations are kept in these relatively narrow concentration ranges by feedback processes between weathering on land, sea level changes, oceanic buffering of CO_2 and biological responses to changes in nutrient and water cycles (e.g. Falkowski et al. 1998, Geider et al. 2001). These regulation mechanisms are not fully understood. Natural variations since the last ice age (11 kyr) have been small, ranging between 260 ppm and 280 ppm and are linked with a relatively stable climate which enabled human civilisation to develop. Present CO_2 concentrations in the atmosphere have increased from 280 ppm of the pre-industrial times (mid of 18th century) to 372 ppm in 2002 and continue to rise at an extremely fast rate of ca. 1.4 ppm/yr (equivalent to 3.2 GtC/yr) compared to all previous geological ages (Figure 3.2.1.1).

These very fast increases are attributed to burning of fossil fuels for energy production (at present 6.3 ± 0.1 GtC/yr) and changes in land use (ca. 1.7 GtC/yr), which to a large extent is driven by deforestation in the tropics (IPCC 2001). On average about half of the anthropogenically released CO_2 is retained in the atmosphere leading to the observed rise since industrialisation. The remaining is taken up to approximately equal amounts by processes on land (1.4 GtC/yr) and in the ocean (1.7 GtC/yr). However, changes in large scale climatic patterns, such as the Southern Oscillation Index have a significant effect on these uptake mechanisms. During El Nino years more CO_2 remains in the atmosphere (Figure 3.2.1.3). These climatic changes affect ocean – atmosphere uptake of CO_2 and primary production on land caused by droughts (Behrenfeld et al. 2001). Similarly, Gruber et al. (2002) estimated from time series observations at Bermuda that climatic effects of the North Atlantic Oscillation may cause an alteration in the strength of the carbon sink in the North Atlantic of ± 0.3 GtC/yr. This natural phenomenon highlights the large influence of climatic patters on the CO_2 fluxes between atmosphere – land – ocean and the large natural variability which has to be taken into account when assessing potential mitigation scenarios.

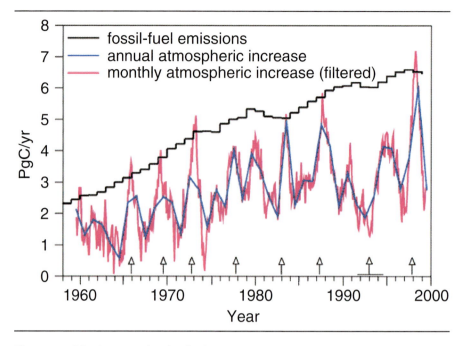

Fig. 3.2.1.3 CO$_2$ Concentration in El Nino years

On longer time scales, the CO$_2$ released by man's activity into the atmosphere can be taken up to 70-80% into the vast carbon storage capacity of the deep ocean (only equilibrium with the atmosphere, not including dissolution of carbonate sediments). However, these time scales span several centuries and do not keep pace with the present rate of change.

Critical Carbon Cycle Processes

A challenge of today's science is to understand the processes which affect the global carbon cycle on time scales relevant to human society. Only some of the natural processes have the right time scales and magnitude to influence the atmospheric rises in CO$_2$ concentration within a few decades and act concurrently with the anthropogenic impacts. These include the biological processes of primary production and respiration, changes in land vegetation, and changes in ocean temperature, stratification and circulation.

Annual plant growth on land and in the sea is a large factor in binding carbon dioxide during the growing season and converting it into organic carbon components. However, of the initially bound carbon (gross primary production = GPP) about half is rapidly being respired again by the plants themselves to CO$_2$ and only the remaining half is transformed to plant tissue (net primary production = NPP). Global terrestrial NPP is estimated as 60 GtC/yr (Saugier and Roy 2001), while global marine NPP is around 45 PgC/yr (Falkowski et al. 1998, Field et al. 1998, Balkanski et al. 1999). Nearly all of this material is being returned to atmospheric CO_2 within a couple of years by the respiration of the organisms living directly or indirectly from the organic carbon fixed by plants. Also combustion of plant biomass by fires returns the organic carbon back to the atmosphere as CO_2. Therefore, the turnover of carbon in this biological recycling system of primary production and respiration is large, but the net balance is close to zero. Only a minor fraction of <1 % of the originally fixed CO_2 is stored in soils or deep-sea sediments for longer time scales. Although this storage is small, accumulation of this small amount over long time scales has led to the gradual decline of atmospheric carbon dioxide concentrations over the millennia and to the deposition of large fossil organic carbon reservoirs.

The important question to be solved is, whether these biological processes can be influenced by changes in environmental conditions and whether the balance between net primary production, respiration and storage in soils and oceanic sediments is altered. Climatic changes (in temperature, radiation, water availability, nutrients and CO_2 concentrations) can affect this balance on land and, partly, also in the ocean. Higher atmospheric CO_2 concentrations can enhance growth of land vegetation as it makes the uptake of CO_2 for photosynthesis easier (Koch and Mooney 1996, Körner 2000). This stimulation of land vegetation is also dependent on the availability of water, nutrients and on temperature. However, in the sea, higher CO_2 concentrations are not likely to stimulate algal growth due to the very high inorganic carbon concentrations already present in the ocean. Plant nutrients, in particular nitrogen sources, have increased both on land and in the ocean due to man's production of fertilisers, combustion of fuels releasing nitrous oxides and wastes from production of animal stocks. The nitrogen compounds are either transported by rivers into the coastal ocean or are released into the atmosphere and later deposited over land or ocean fertilising remote regions. Since most of the plant growth in the open ocean is limited by nitrogenous nutrients such input stimulates marine production and, hence, biological carbon fixation. Also part of the presently observed higher plant production on land was attributed to this fertilisation. Another fertilisation effect comes from dust exported from land and deposited over the ocean. This dust export is partly influenced by agriculture and originates mainly from disturbed soils (Tegen and Fung 1995). Dust is a transport mechanism for trace elements, such as iron, which is deficient in some large oceanic regions remote from land, such as the Southern Ocean, tropical Pacific and the North Pacific. Large scale experiments in which iron was artificially added in such ocean regions to the surface water stimulated algal growth and proved that iron is indeed a limiting element for ocean primary production (e.g. Boyd et al. 2000). Evidence from ice core records indicates that dust flux and CO_2 concentrations fluctuated in synchrony highlighting a possible chain of causal links: changes in vegetation cover on land affect dust export to the open ocean, this stimulates primary production in iron-deficient regions and leads to changes in CO_2 uptake by such biological processes in the ocean. Although in some ocean regions such alteration of productivity is likely, the resulting carbon fluxes are not well quantified and most assessments assume that the global biological carbon uptake processes in the ocean will remain largely unchanged in the near future.

A significant impact on carbon fluxes is related to man made changes in vegetation cover by agriculture and forestry (1.7. GtC/yr). In particular, deforestation in the tropics and desertification is responsible for substantial release of CO_2 into the atmosphere. Efforts are underway to counteract this release of CO_2 by reforestation and changes in land management practices to enhance the terrestrial carbon sink. The magnitude of this sink is estimated as −3.8 to +0.3 GtC/yr (av. −1.9 GtC/yr), but it is still poorly defined (Prentice et al. 2001). Changes in land use and reforestation are one of the options discussed for management of carbon fluxes since they can be directly influenced by man and are of a significant magnitude. The largest capacity for such management exists in the tropics and in the extra-tropical Northern Hemisphere lands. To use forests as long term carbon storage, forests have to be replanted when harvested, the harvested wood has to be utilised for long-term products or for energy production from biomass to reduce the consumption of fossil fuels. It has been estimated that such practices could increase the terrestrial carbon stocks by 60 to 87 GtC in the period from 1995 to 2050 (Brown et al. 1996). A large uncertainty is the storage of carbon in soils, which is less effective in managed forests and in tilled land. Changes in agricultural practices (fertilisation, irrigation, residue management, reduced tillage) can increase the carbon storage in soils. The potential for net carbon storage of changes in land use has been estimated as 10 to 30% of total anthropogenic emissions (Prentice et al. 2001). However, this storage capacity is limited to an upper boundary of 40 to 70 ppm over the next century. In contrast, continued deforestation has a much larger potential to increase atmospheric CO_2 concentrations by a factor of 2 to 4.

Physical and chemical processes dominate the oceanic CO_2 fluxes. The increasing atmospheric concentrations cause a higher diffusion of CO_2 into the surface ocean. However, since the solubility of CO_2 in sea water declines when temperature rises, warming of the surface ocean will lead to decreased uptake of CO_2. The surface ocean is characterised by multidecadal temperature oscillations. At present, warming in many regions of the ocean has accelerated over the last five years. The buffering capacity of the ocean carbonate system will become less effective at higher CO_2 concentrations which means that the oceanic sink for CO_2 will become smaller in the future. Increasing temperatures as well as increasing freshwater influx in the high latitudes will also strengthen the stratification of the upper ocean. The general effect of this is that the surface waters are more strongly separated from the deep waters and that vertical transport processes are reduced. Under these conditions biological production in the tropical and temperate parts of the ocean tends to decline due to reduced upward flux of nutrients. In the high latitudes, reduction of vertical mixing and deep water formation is a likely consequence resulting in less CO_2 uptake by the ocean. At the same time, in the higher latitudes increased stratification will enhance biological production, because shallower stratification provides a better light climate for the plankton algae. The net result of these counteracting processes is difficult to estimate, but it is assumed that the physical processes dominate carbon fluxes and that they are counteracted by biological processes to a significant degree. Due to the very large area covered by the Southern Ocean the biggest effects are suspected to occur at latitudes between 40°S and 70°S (Sarmiento et al. 1998). These changes can substantially affect the ocean carbon sink in the next few decades. The largest unknown in these estimations is the reaction of the biological processes to the altered physico-chemical environment in the different ocean regions. It is assumed that under future scenarios of higher CO_2 concentrations and warmer temperatures the ocean will become a less effective carbon sink than in the past.

At present several options are discussed to intentionally manipulate carbon fluxes with the purpose to increase the uptake of carbon into the ocean. One possible manipulation of carbon fluxes in the sea is the intentional fertilisation of specific ocean regions with iron to stimulate algal production. However, the effectiveness of this procedure in respect to carbon storage is disputed and undesired side effects are certain (Chisholm et al. 2001). A second manipulation is the "dumping" of CO_2 into the deep ocean in form of CO_2–enriched water, liquid CO_2 or similar condensed forms. The time of storage depends on the depth into which the CO_2 is deposited and can range from a decade, when released in depths of a few hundred meters, to centuries, when deposited in more than 2000 m depth. Increasing pH at the sites of CO_2 introduction may have serious effects on the biota and on geochemical processes, but this is at present poorly understood and, hence, the associated impacts on the marine ecosystem are difficult to assess. In a similar way, CO_2 can be deposited in geological formations, such as exhausted gas reservoirs. This is already tested on industrial scale and is likely to offer a less harmful way of CO_2 storage for the long term.

Research issues

First insights into the variability of carbon fluxes and their regulation have been drawn from long-term observations as well as assessments of global carbon stocks and fluxes. Therefore, future research is heavily based on extended and well coordinated observational programmes. This will be carried out by space agencies as well as environmental monitoring programmes and is part of the research agendas of the atmospheric, terrestrial and marine IGBP and WCRP programmes. In order to improve this research, the boundaries between basic research and monitoring, that exist both in respect to scientific attitude as well as in funding have to be overcome.

A second valuable tool are advanced models which couple physical, chemical and biological processes. Much progress has been achieved already in this respect, but much remains to be done as well. The challenge lies in understanding the various interlinked feedback processes that affect different elemental cycles and all parts of the Earth system. A particular gap is the missing connection between models describing natural processes and models of economic and social behaviour. This requires completely new approaches as in these fields of science different model philosophies prevail and the focus differs between global versus regional processes.

Despite a decade or more of research into the processes controlling carbon fluxes much basic research still needs to be done. This concerns, in the first place, the links between the different components of the earth system and the explicit consideration of human action as part of the carbon cycle. At the same time, pressing questions in respect to potential regulation of anthropogenic carbon fluxes have to be answered. This requires to focus research on the mechanisms that could realistically lead to stabilisation of atmospheric CO_2 concentrations. Natural processes, technological feasibility, economic consequences, legal considerations and political instruments need to be considered in a comprehensive approach. This will require an exceptional degree of joint research between these fields of science which has not been achieved in the past. Aspects of such joint research are discussed in the following three chapters.

Scientific foundations for carbon sink spatial assessments and management

The Kyoto Protocol and subsequent discussions raised the issue of offsetting emissions by deliberate management of carbon sinks. A fundamental question for management of atmospheric carbon levels is: "Are there large carbon flows within the carbon cycle that humans can deliberately and significantly influence?" A pre-requisite for answering this question is to identify significant carbon flow pathways, together with their geographical location and extent, and the mechanisms that drive and control them. Assuming potentially manageable pathways are identified, the next logical questions are: "What techniques can we use to influence these flows?" and "How much would this management cost, how effective would such management be and what are the associated risks?".

Basic research into this area is required in order to provide a foundation from which to assess the credibility and effectiveness of proposed activities. Techniques to assess carbon flows range from the operation of eddy-correlation measurements of CO_2 fluxes from tall 'flux towers' on local to regional scales, through to inverse modelling of net fluxes on regional to continental scales. For assessing these larger scales through the model-based interpretation of atmospheric CO_2 data, the flux of carbon between the atmosphere and the ocean, and the large-scale patterns of CO_2 in the atmosphere, become critical information. Ship-based measurements of surface water CO_2 concentrations together with state-of-the-art models of the upper ocean can potentially be used to infer three-dimensional fields of air-sea fluxes. New satellite-based sensors can be used to infer spatial distributions of atmospheric CO_2 over the oceans and over the continents. Because of the complexity of carbon flows within the atmosphere-ocean-land system and our inability to measure flows at all locations at all times, measurements must always be combined with models to fill the gaps and cross-check or compare observations against expectations that are based on mechanistic understanding.

Given the complexity of carbon flows, the associated measurement difficulties, and the fact that the key flow pathways of interest are between atmosphere, land and ocean, it is worthwhile to integrate the information concerning inter-reservoir fluxes and atmospheric distributions. The atmosphere serves as an integrator of carbon flows between ocean and atmosphere and between land and the atmosphere. Hence it can be that measurement of the (relatively) slow-varying and geographically homogeneous flows of carbon between atmosphere and ocean may provide useful information to allow better characterisation of the more temporally and spatially variable fluxes between land and atmosphere.

At present, the various capabilities listed above are not fully integrated. In particular, the link between ocean-atmosphere and land-atmosphere flux estimations has not been firmly established either scientifically or institutionally. Similarly, the potentially unique power of satellite-based trace gas measurements for revealing carbon-flow distributions has not been fully integrated with the other sources of information and models.

A scientific framework and a set of data handling and data interpretation methodologies are required which will allow diverse sources of information concerning carbon fluxes and spatial distributions to be combined, interpolated in space and time, assessed and interpreted. Models are a critical component of this framework for both the interpolation (e.g. data assimilation models) and interpretation (e.g. mechanistic models of the terrestrial biosphere).

The output of such a framework will be the knowledge, techniques and infrastructure required for observation-based spatial characterisation of carbon flows between the atmosphere, ocean and land surface. This spatial information in turn can be directly compared with both:

- Mechanistic models of terrestrial carbon cycling involving soils, agriculture and forest dynamics
- Spatial distributions of economic activity including, but not limited to, industrial CO_2 emissions

The goal of such a spatial comparison of fluxes and mechanisms is to reveal the distribution, scale and underlying causes of carbon flows and, hence, permit a realistic assessment of their potential for management.

Pathways to decarbonisation

Reduction of carbon emissions is a potential way towards stabilisation of atmospheric CO_2 concentrations. This is directly linked to technical developments, economy and changes in life styles of societies. It has to be determined which pathways to a sustainable, non-carbon-fuelled energy system are biophysically, socially and economically possible. This also entails the question, which mix of adaptation to and mitigation of climate change is optimal for society and for the ecosystem. To begin to answer these questions will require fundamental research that cuts across the boundaries between biophysical and human sciences. The following major topics emerge.

- The conditions for, and consequences of, different approaches to emissions reduction. The approaches include multiple non-carbon energy sources, biofuels, improvements in energy efficiency, and lifestyle changes associated with consumer choices and the trend towards an "information society". For each approach, there needs to be understood and quantified its technological feasibility, economic viability, social acceptability, effectiveness (in terms of its impact on emissions), and collateral impacts such as effects on other aspects of the environment.
- Analogously, we need to understand and quantify the same variables for purposeful methodologies to sequester carbon or to preserve existing carbon stocks. These methodologies include afforestation and reforestation, changing agricultural and forest management practices, iron fertilisation of the ocean, deep-sea dumping of CO_2, and developing incentives to prevent loss of extant forests.
- As some degree of climate change is now an inevitable concomitant of any scenario of the future, and as assessment of different scenarios will require comparison of different degrees of climate change impact on the environment and society, it will be necessary to further develop the science of climate impacts to be able to analyse the feedbacks from climate change, via various economic and resource sectors, on population and economic growth.

Each of these topics calls for a mixture of different kinds of research, including:

- Fundamental observational research and numerical modelling: for example, flux measurements to quantify carbons uptake by new forests and models to analyse the consequences of forestation scenarios for the global carbon balance
- Data synthesis, especially linking aspects of the biophysical and socio-economic systems (e.g. quantifying energy consumption associated with different lifestyles)
- Improved integrated assessment modelling, including an increased emphasis on techniques for model evaluation by analogy with the historical approach commonly used to test biophysical models
- Involvement of stakeholders in the design of key indicators and the design of models

A major intermediate objective of this research will be the development of "post-SRES" (IPCC Special Report on Emission Scenarios) emissions scenarios in which climate policy instruments are explicitly included. It will then be possible to begin to explore the multidimensional space of possible joint trajectories of climate, human dimensions and energy systems and seek for trajectories fulfilling the sustainability requirement.

Human Society and the Carbon Cycle

Since humans have become a driving force of similar magnitude as natural forces in respect to the carbon cycle in the last 200 years, man's role in the global carbon cycle, both as object and subject of change, has to be a central part of research. While the impact of human existence on the carbon cycle may have been unintended, its reduction, however, will require an intentional global effort on an unprecedented scale.

Man's perception of the carbon cycle, apart from a small community of scientists and politicians, is characterised by widespread misconceptions and lack of understanding. This is not surprising considering the complex nature of the problem and the many uncertainties. If carbon policies are to be met by public approval, the general understanding of the carbon cycle and of its role in global change has to be improved guided by questions of "How to teach people grasp complex systems such as the carbon cycle? What are the interests and perceptions of risk that shape man's view of the world and how might they be changed? What are the value judgements inherent in taking decisions and changing people's behaviour?" These basic concepts and preconceptions about the carbon cycle need to be investigated in different societies.

The instruments that could help to meet the challenge posed by perturbations of the carbon cycle have to be evaluated. The question is, whether the future challenges of the anthropocene age require new tools. The relatively new concept of "emissions trading" has to be analysed in respect to its effectiveness and consequences. This instrument is characterised by a strong combination of ecological effectiveness and economic efficiency. Since it is an economic instrument based on quantified targets to reduce carbon emissions (and potentially other greenhouse gases), it connects the natural and the social sphere. Emissions trading might thus serve as a link between the ecological necessities posed by the perturbations of the carbon cycle and the societal needs for efficient allocation. Measures to take effective control over the human induced carbon fluxes may be in conflict with other desirable aims, such as protection of biodiversity or food production. These potential conflicts have to be addressed in a whole-system approach.

This leads to the question whether the anthropocene requires new forms of governance in order to meet the challenges of global change. If management of the global carbon cycle becomes a necessity, what kind of governance or government would best serve this goal without compromising values such as personal liberty and democracy? What is the appropriate balance between control and freedom? How to ensure equity between different countries, especially between developed and developing countries? The Kyoto Protocol, for the first time in history, has made a deliberate attempt at the international level to decouple economic growth and the emissions of greenhouse gases. Therefore, it is important to critically assess the legal, economic and governance tools which are available to society to address the perceived problems of rising CO_2 concentrations in the atmosphere.

Concluding remarks

The global carbon cycle has attained a high political and public visibility. We fear that the rapid rise in atmospheric CO_2 concentrations, which has no parallel in the past history of earth, may trigger climatic and environmental changes. This cannot be seen in isolation, but feedbacks with other processes, such as the hydrological cycle and aerosols, are likely to cause substantial environmental changes. In combination with our presently very high human population, which is occupying more and more marginal habitats, this may lead to dangerous future conditions for part of Earth's human population.

Recognising this development, research has to address the questions "What is the right mix of adaptation or mitigation to cope with future changes? Which parts of the carbon cycle can be managed and what are the consequences?" Future research, therefore, has to develop a whole-system perspective which encompasses natural and socio-economic sciences. This research requires:

- joint collection of data of natural, economic and social variables;
- identification of the mechanisms which are likely to change carbon fluxes on time scales relevant for society;
- identification of mechanisms amenable to human control and their effectiveness in respect to stabilisation of atmospheric carbon concentration;
- consideration of natural, technical, economic, social, legal and institutional aspects in models and assessments;

Since model forecasts of a range of different carbon scenarios predict that significant changes in the carbon cycle and in the climate system are likely to manifest themselves after about 50 years from now, society is asked to develop visions and plans of action now. This requires a new area of highly interdisciplinary and targeted research.

Acknowledgements

This article is based on the discussions of two "Carbon -Workshops" held in Kiel sponsored by the DFG. I am grateful to the participants of the workshop for their valuable inputs: Horst Bayrhuber, Jelle Bijma, Claus Böning, Harald Bradke, Michael Buchwitz, Ottmar Edenhofer, Gernot Klepper, Arne Körtzinger, Corinne LeQuéré, Margit Lott, Eckhard Lucius, Fritz Reusswig, Monika Rhein, Ulf Riebesell, Hermann E. Ott, Colin Prentice, Jörg Priess, Ralf Schuele, Victor Smetacek, Richard S.J. Tol, Christiane Trüe, Douglas Wallace and Pep Canadell (GCP Executive Officer, Canberra Australia).

References

Balkanski YP et al. (1999) Ocean primary production derived from satellite data: an evaluation with atmospheric oxygen measurements. Global Biogeochemical Cycles 13:257-271

Behrenfeld MJ et al. (2001) Biospheric primary production during an ENSO transition. Science 291:2594-2597

Berner RA (1997) The rise of plants and their effect on weathering and atmospheric CO2. Science 276:544-546

Brown et al. (1996) Management of forests for mitigation of greenhouse gas emissions. In: Watson RT et al. (eds) IPCC Climate Change 1995 – Impacts, adaptations and mitigation of climate change: scientific-technical analyses, contribution of working group II to the second assessment report of the Intergovernmental Panel on Climate Change. Cambridge University Press, Cambridge, pp 773-797

Boyd PW et al. (2000) A mesoscale phytoplankton bloom in the polar Southern Ocean stimulated by iron fertilization. Nature 407:695-702

Falkowski PG, Barber RT, Smetacek V (1998) Biogeochemical controls and feedbacks on ocean primary production. Science 281:200-206

Field CB et al. (1998) Primary production of the biosphere: integrating terrestrial and oceanic components. Science 281:237-240

Geider RJ et al. (2001) Primary productivity of planet earth: biological determinants and physical constraints in terrestrial and aquatic habitats. Global Change Biology 7:849-882

Gruber N, Keeling CD, Bates NR (2002) Interannual variability in the North Atlantic Ocean carbon sink. Science 298:2374-2378

Hayes JM et al. (1999) The abundance of 13C in marine organic matter and isotopic fractionation in the global biogeochemical cycle of carbon during the past 800 ma. Chem Geol 161:103-125

Houghton JT et al (eds) (2001) Climate Change 2001: the scientific basis. Contribution of Working Group I to the Third Assessment Report of the Intergovernmental Panel on Climate Change. Cambridge University Press, Cambridge

Kleypas JA et al. (1999) Geochemical consequences of increased atmospheric carbon dioxide on coral reefs. Science 284:118-120

Koch GW, Mooney HA (1996) Response of terrestrial ecosystems to to elevated CO2: a synthesis and summary. In: Koch GW, Mooney HA (eds) Carbon dioxide and terrestrial ecosystems. Academic Press, San Diego, pp 415-429

Körner C (2000) Biosphere responses to CO2 enrichment. Ecological Applications 10:1590-1619

Pearson PN, Palmer MR (2000) Atmospheric carbon dioxide concentrations over the past 600 million years. Nature 406:695-699

Penner JE et al. (2001) Aerosols, their direct and indirect effects. In: Houghton JT et al. (eds) Climate Change 2001: The Scientific Basis. Contribution of Working Group I to the Third Assessment Report of the Intergovernmental Panel on Climate Change. Cambridge University Press, Cambridge, pp 289-348

Petit JR et al. (1999) Climate and atmospheric history of the past 420,000 years from the Vostock ice core, Antarctica. Nature 399:429-436

Prather M et al. (2001) Atmospheric Chemistry and greenhouse gases. In: Houghton JT et al. (eds) Climate Change 2001: The Scientific Basis. Contribution of Working Group I to the Third Assessment Report of the Intergovernmental Panel on Climate Change. Cambridge University Press, Cambridge, pp 239-288

Prentice IC et al. (2001) The carbon cycle and atmospheric carbon dioxide. In: Houghton JT et al. (eds) Climate Change 2001: The Scientific Basis. Contribution of Working Group I to the Third Assessment Report of the Intergovernmental Panel on Climate Change. Cambridge University Press, Cambridge, pp 183-238

Riebesell U et al. (2000) Reduced calcification of marine plankton in response to increased atmospheric CO2. Nature 407:364-367

Sarmiento JL et al. (1998) Simulated response of the ocean carbon cycle to anthropogenic climate warming. Nature 393:245-249

Saugier B, Roy J (2001) Estimations of global terrestrial productivity: converging towards a single number? In: Roy J, Saugier B, Mooney HA (eds) Global terrestrial productivity: past, present and future. Academic Press, San Diego

Tegen I, Fung I (1995) Contribution to the atmospheric mineral aerosol load from land surface modification. Journal of Geophysical Research 100(D9):18707-18726

3.2.2 International Institutions and the Carbon Regime

Gernot Klepper

Kiel Institute for World Economics, Düsternbrooker Weg 120, 24105 Kiel

Introduction

The United Nations Framework Convention on Climate Change (UNFCCC) starts out with the acknowledgement that climate change and its adverse effects are a concern of humankind thus requiring an international treaty aimed at "the widest possible cooperation by all countries and their participation in an effective and appropriate international response" (UNFCCC 1992). Although Art. 2 which states the major objective of the UNFCCC refers to a stabilisation of the concentration of greenhouse gases in the atmosphere it may sound as if only atmospheric aspects of the climate problem were addressed. Already the preamble, however, makes clear that the role of terrestrial and marine ecosystems as sinks and reservoirs of greenhouse gases are important.

The Kyoto-Protocol to the UNFCCC also recognises these interactions by including greenhouse gas emissions as well as sinks in the commitments that the industrialised countries should take in order to take a first step towards achieving the objective of Art. 2 UNFCCC. The Kyoto Protocol has not come into force since the United States have decided not to sign it and Russia is still discussing her signature which would render the Protocol into force. Nevertheless, with the Kyoto-Protocol coming into force or failing, a process has begun which establishes international and national institutions for the management of the carbon cycle. This process has surely not covered the whole of the carbon cycle, but atmospheric emissions are subject to controls in most countries.

The research on atmospheric carbon flows has directed the attention towards the complete carbon cycle since there exist important interactions between the atmospheric, terrestrial, and marine carbon flows. The process of trying to understand these interactions is still under way. As a better knowledge about the complete carbon cycle emerges, the question arises as to whether this knowledge can and should be incorporated into the climate policy process and subsequently into the design of international institutions and conventions, which coordinate policy instruments for managing the carbon cycle.

In the following a very brief description of the carbon cycle is introduced in order to identify the origins of human interference and to locate possible ways to manage the carbon cycle. This is followed by a short discussion of issues concerning an optimal management of the carbon cycle. From this "nirvana approach" practical steps are derived for the management of the carbon cycle under real world conditions, i.e. given institutional and political constraints but also constraints on the knowledge about the carbon cycle. Finally, some conclusions are drawn.

The Carbon Cycle

Carbon occurs in all compartments of the earth, on land, in geological reservoirs, in the atmosphere, and in the oceans. Huge amounts are stored, but there is also a permanent exchange between the biosphere on land, the atmosphere, and the ocean. This transport of carbon across the media is called the "carbon cycle".

The carbon cycle by definition is always balanced, i.e. total net flows are zero. However, the magnitudes of single flows can change. The global estimates of the carbon cycle by the IPCC are represented in Figure 3.2.2.1. The atmosphere con-

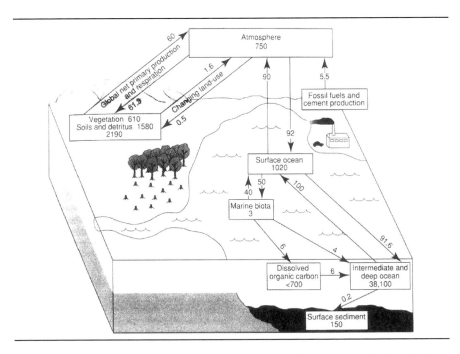

Fig. 3.2.2.1 Main Components of the Natural Carbon Cycle Reservoirs (GtC) and Fluxes (GtC/yr)

taining a stock of about 750 GtC receives about 60 GtC from soils and plants and about 90 GtC from the oceans, but it also returns slightly more than that. These fluxes are subject to substantial natural variations; yet they are roughly balanced.

This carbon cycle has been disturbed by human activity since fossil fuels have been taken from the geological stock and burned, thus adding to the carbon flows into the atmosphere. An additional human contribution comes from the change in land use and land cover by humans. Through land management the amount of carbon stored in soils and in biomass can be changed such that carbon flows from land to atmosphere and ocean can increase or decrease. Both activities, therefore, mobilise of immobilise carbon from the stocks and thus change the carbon cycle. Figure 3.2.2.2 illustrates the human perturbation of the carbon cycle.

Comparing carbon stocks, carbon flows, and the human influence on carbon flows reveals a clear difference in orders of magnitude. Ignoring geological reservoirs, carbon stocks on land are composed of carbon in soil (1500 GtC) and plants (500 GtC). These 2000 GtC and the 730 GtC in the atmosphere are small compared to the 38.000 Gt of carbon in the oceans. However, only a small percentage of these stocks is actually exchanged between the media. The atmosphere receives the largest share – compared to its stock – with 60 GtC from land and 90 Gt from the ocean. The land has a very small transport to the ocean but the yearly flows to the atmosphere amount to 120 GtC compared to its stock of about 2000 GtC.

The human contribution to these flows is even smaller. The addition of fossil fuels from the geological reservoir to the carbon cycle was between 5 and 6 GtC

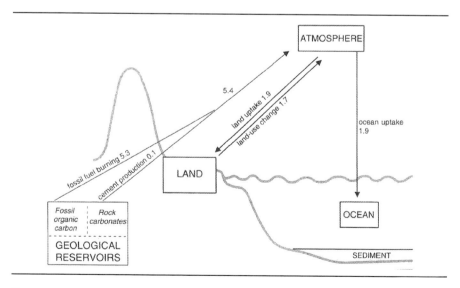

Fig. 3.2.2.2 Human Influence of the Carbon Cycle Source: IPCC (2001a)

per year in the 1980s and amounts to 6-7 GtC in the 1990s, i.e. roughly 1 percent of the atmospheric stock per year. Soil and plants receive 1.9 GtC from the atmosphere and release through land-use change about 1.7 GtC, hence there is a small net transfer to land, i.e. the carbon stock on land is slowly increasing at the moment. Finally, the oceans take up about 1.9 GtC. As a consequence, the human emissions of carbon through burning fossil fuels (about 6.3 GtC, including cement production) and through land use changes (about 1.7 GtC) increase the carbon in the atmosphere by about 8 GtC. However, the carbon cycle adjusts to this perturbation by moving about 1.7 Gt of carbon into the ocean. Also a fraction of the emissions from land are returned. Table 3.2.2.1 summarises the state of knowledge from the Third Assessment Report of the IPCC.

As Table 3.2.2.1 indicates, roughly 8 GtC are added to the stock of carbon in the atmosphere every year but only between 3 and 4 GtC remain there. The difference is taken up by the ocean and by the so-called "residual terrestrial sink". These terrestrial sinks are not well understood and are only measured indirectly by comparing the deposition in the air with the net change in the carbon stock in the atmosphere. The search for this "missing carbon" is still under way. The main factors which contribute to the up-taking of the carbon from the atmosphere is the re-growth of forests in the northern hemisphere and an overall fertilisation effect, both through CO_2-fertilisation and human induced atmospheric deposition of Nitrogen. However, model calculations show that the residual sink capacity will diminish and the residual terrestrial sink will fall. Hence, it will become increasingly difficult to control the CO_2-concentration in the atmosphere by hoping that a good proportion of the fossil fuel emissions will go back to marine or terrestrial sinks (Noble 2001).

It is also evident that the uncertainties about the size of the terrestrial and the marine sinks are significant. E.g., it is not clear whether in the 1980s the land was a net sink or a net emitter of carbon. In addition, the IPCC seems confident that in the 1990s land was indeed a net sink but the reasons and causes for these fluxes are not known so far.

Table 3.2.2.1 Global CO_2-Budgets (in GtC)

	1980s	1990s
Atmospheric increase	3.3 ± 0.1	3.2 ± 0.1
Emissions (fossil fuel, cement)	5.4 ± 0.3	6.3 ± 0.4
Ocean-atmosphere flux	−1.9 ± 0.6	−1.7 ± 0.5
Land-atmosphere flux	−0.2 ± 0.7	−1.4 ± 0.7
Of which:		
Land-use change	1.7 (0.6 to 2.5)	NA
"Residual terrestric sink"	−1.9 (−3.8 to 0.3)	NA

Source: IPCC (2001a)

Managing the Carbon Cycle

The global carbon cycle is undergoing a slow but ongoing transformation through the impact of human activity. This can most notably be observed by the increase of CO_2 in the atmosphere from close to 280 ppm in 1800 to a value of 367 ppm in 1999. This increase will continue and reach a value of between 550 ppm and almost 1000 ppm at the end of this century according to the different scenarios of the IPCC (2001) unless some measures are taken to limit such an increase.

Since at least the higher values of the IPCC scenarios seem quite unacceptable, it is clear that the carbon cycle requires some management. There are in principle two ways in which this can be done. As shown above, human activity increases the carbon flows by mobilising geological reservoirs – predominantly by burning fossil fuels – and by mobilising carbon stored in soils and in plants through purposeful changes in land-use, both for agricultural production and for other human uses of land. These direct influences can be managed by controlling the use of fossil energy and by controlling land-use changes.

The other means of managing the carbon cycle consists of a direct interference in the carbon flows between the carbon stocks in the ocean, on land, in geological reservoirs, and in the atmosphere, i.e. it is an attempt to control the repercussions of the mobilisation of carbon once they have occurred. The list of management options in this area is long consisting of many sequestration options:

- *Land-use changes* can affect the carbon reservoir in soils and in plants. Especially forests and the use of forest products may play a vital role in removing atmospheric CO_2 and storing it on land.
- *Carbon sequestration in the ocean* is an activity that is intensively researched. The idea is to induce the marine biosphere to take up more carbon that then will be moved into the large reservoirs of dissolved inorganic carbon. This 'biological pump' is not well understood so far, especially in terms of its ecologic side effects.
- *Technological options* for carbon capturing in industrial processes and storage in geological reservoirs are also an option that is in an early stage of applied research.

A comprehensive management strategy for the carbon cycle would need to evaluate all these options in the effort to reach a certain objective, e.g. a stabilisation of CO_2- or greenhouse gas concentration at a certain level as proposed in the Framework UNFCCC. Ideally, this process would consist of assembling a comprehensive list of mitigation options with their contribution to the control of the carbon flows. It would for each option identify the side effects in terms of impacts on objectives other than climate mitigation. It would establish the repercussions on the carbon cycle itself and on the adaptive capacity of the different carbon stocks. Finally, the social cost of each management option would need to be assessed.

Based on this assessment a theoretically optimal management of the carbon cycle could be computed. It would consist of an emission path for carbon from fossil fuels, a regionally disaggregated land-use strategy, and a sequestration strategy into terrestrial and marine carbon reservoirs. Such program would be cost minimising economically and it could limit negative side effects as much as possible. However, its main disadvantage is: It cannot be implemented! The three most important reasons are:

- There is not enough knowledge about the details of the carbon cycle,
- there does not exist a manager who could implement the optimal carbon regime, and
- the administrative burden of such a regime would be overwhelming.

A carbon regime, therefore, has to cope with all three deficiencies. A lack of information can slowly be overcome to some extent by research, yet it has to be recognised as long as it exists. This issue will not be further discussed here. Instead, the focus will be on the other two obstacles, the lack of a worldwide authority and the need for simple solutions that can be implemented in practice.

Implementing a Carbon Regime

Implementing a carbon management regime builds on an already existing structure of local, national, and international policies and institutions. On a global scale the large majority of states have signed the UNFCCC, which states in Art. 2 that the international community should seek to stabilise GHG-concentrations in the atmosphere at a level that would prevent dangerous anthropogenic interference with the climate system. This should be achieved within such a time frame that ecosystems could adapt naturally to climate change, that food production is not threatened, and that economic development can proceed in a sustainable manner. Interpreting this common goal is a difficult process, which is currently under way and has, of course, important implications for the extent of interference in the current carbon cycle (Ott et al. 2003, Klepper et al. 2003).

The Kyoto Protocol – which has not come into force yet – together with its interpretation by the parties to the protocol during the different COPs defines one specific institutional setting for the control of a part of the carbon cycle. It defines national commitments for the Annex I countries on the emission of GHG control by the year 2008 to 2012. It also includes the possibility of enhancing terrestrial sinks. The longer term "Post-Kyoto-Process" is not determined yet.

In conjunction with the Kyoto-Protocol many national governments have introduced a more or less comprehensive set of policies that control emissions of GHGs, most prominently those of CO_2. These policies are often supplemented by local initiatives within the framework of the Agenda 2000 process.

These activities on different levels of decision making reflect a particular feature of the carbon management. The human interference with the carbon cycle usually takes place at the individual decision level of the household or the firm. These entities decide about the emissions for CO_2 and about land-use changes. These decisions can be influenced only by state authorities and by public opinion. However, the effects of such interference with the carbon cycle – be it with a positive or a negative impact – occurs worldwide. This is the classical case of a global externality where the benefits of using carbon accrue to the individual that uses carbon, and the costs in terms of climate change are borne by the world community. In addition, these costs are often distributed unevenly such that those in the North who consume most of the fossil energy often are not as negatively affected by climate change as those in the South who only contribute a small share to the world wide CO_2-emissions (IPCC 2001b).

The global externality problem applies to most of the human interference in the carbon cycle. The addition of carbon to the carbon cycle through the burning of fossil fuels and through the mobilisation of the terrestrial carbon stock from land-use changes is an individual activity with global consequences. The reverse interference, i.e. the removal of carbon from the atmosphere as demanded by Art. 2 UNFCCC also is an individual activity since either the land owner needs to increase the carbon stock on his land or the firm needs to capture carbon from its industrial processes and return it to geological or marine reservoirs. These activities do not constitute negative externalities as the emissions mentioned above but positive externalities. For positive externalities the individual actor bears the costs and the global community reaps the benefits. Hence, in both cases benefits and costs do not accrue to the same decision maker. One special case that includes an additional complication is the case of carbon sequestration on international waters. This would need to be done by institutions, which are under some national jurisdiction, but their activity takes place on international waters where the legal aspects for such activities are far from clear. This special case will be discussed below.

The presence of negative or positive externalities presents an economic allocation problem. Under negative externalities an individual has an incentive to mobilise more carbon than would be optimal from a global point of view. This is the case since that individual does not face the costs of his activity. This phenomenon is often called the "free rider problem". The conflict between private decisions and societal goals can only be resolved by state action that regulates the individual behavior in such a way that it confirms with the objectives of the society.

The same reasoning can be applied to the positive externalities created by activities such as land use changes that capture or carbon sequestration in industrial processes. Individual actors as well as governments do not have an incentive to engage in such activities as long as they are not compensated for their efforts by those who benefit from these activities. Therefore, in the same way as the negative externalities induce higher than optimal carbon flows into the atmosphere, positive externalities induce too little creation of sinks for atmospheric carbon.

The externality problem becomes even more complex in the case of global externalities. National governments like individual agents have an incentive to free ride on other nations by not taking into account the external costs that their citizens impose upon the world community. Sovereign states are generally believed to only act in the interest of their own citizens and not on behalf of foreign citizens. Since there does not exist a world government that has the power to regulate national regulatory action global externalities are far more difficult to control than externalities that are confined to occur within state borders.

An efficient and effective carbon management can only be achieved if the international incentive problems are resolved. In the current international order this can happen through international environmental agreements in which sovereign states agree on a joint effort to a specific carbon management regime. Such agreements, however, are difficult to achieve and difficult to maintain (Heister 1997, Barrett 2003). In fact, there are only a few international environmental agreements that turned out to work sufficiently effective. A prime example is the Montreal Protocol for controlling substances that destroy the ozone layer.

As far as the carbon cycle is concerned, the UNFCCC does not contain any definite commitment by the parties, and the Kyoto Protocol has not come into force so far.[1] Nevertheless, the Kyoto Protocol contains some controls of the carbon cycle, but it also reveals a number of serious weaknesses that arise as a carbon management regime is to be established within an international agreement.

The Kyoto Protocol signed in 1997 so far is the only international institution that deals with the carbon cycle. It covers only a part of the carbon cycle namely it is only concerned with the GHG-emissions of Annex I-countries. It also sets limits to emissions by country, hence it controls the carbon flow from the use of fossil fuels. The emission caps applied to Annex I-countries cover only about 60 percent of world emissions. Since the Kyoto targets are defined in GHGs, these targets also allow a substitution of CO_2 by other GHGs. In addition, a country may – according to Art. 3.3 of the Kyoto-Protocol – use an enhancement of sinks to lower its emission cap. Hence, the two arrows in Figure 2 marked "fossil fuel burning" and "cement production" are not separately controlled, but a country can substitute this arrow with the arrow in the opposite direction, namely "land uptake".

The articles 3.3 and 3.4 of the Kyoto-Protocol on land-use change allow to include in the cap "direct human induced land-use change and forestry activities, limited to afforestation, reforestation and deforestation since 1990" (Art. 3.3). However, Art. 3.4 leaves it to the parties of the Protocol to decide which activities in the area of land-use, land-use change, and forestry (LULUCF) should be included in the second commitment period. These human induced changes in the carbon stock in plants and soils raise many complicated measurement, verification, and benchmarking issues (Watson, Noble 2001).

The Kyoto Protocol as the only comprehensive institutional set up for a carbon management regime controls only a part of the global carbon cycle. It has specific provisions – although surely not far reaching enough – on the emissions from burn-

---

[1] With Russia signing the Kyoto-Protocol reaches treaty status.

ing fossil fuels, and it raises several management questions concerning terrestrial sinks that should be resolved by a concerted effort of the parties to the Kyoto-Protocol. This raises the question as to whether a carbon regime should be extended beyond the coverage of the Kyoto Protocol, whether it can be done, and if so, how it should be done.

From the Kyoto Protocol to Global Carbon Management

The question as to whether the carbon control should be extended beyond the scope of the Kyoto-Protocol at first depends on the quantitative question if the components of the carbon cycle that can be influenced by human activity and which are not covered are important from a carbon control point of view. Then one needs to find out if it is worth doing so in terms of the welfare to the global community.

As Table 3.2.2.1 clearly shows although the emissions of CO_2 from fossil fuel burning currently amount to 6 to 7 GtC per year only somewhat more than 3 GtC remain in the atmosphere thus causing the disequilibrium in the climate system. Close to 2 GtC are fluxes into the ocean, so the terrestrial sink has absorbed in the 1990s between 1 and 2 GtC. This is obviously not a negligible amount. The overall potential of LULUCF activities for the next 50 years may even be in the order of 100 GtC accumulated over this period. This is 10 to 20 percent of the projected CO_2-emissions from fossil fuels. Simulations with 6 dynamic global vegetation models also show that the terrestrial sink capacity will more then double until about 2030. Unfortunately, after that time the sink will start to fall as terrestrial carbon storage capacities seem to become exhausted (Figure 3.2.2.3) (Cramer et al. 2001). These results indicate that LULUCF activities may have an important role to play in the medium run for balancing the so far relatively unsuccessful attempts in controlling fossil CO_2-emissions.

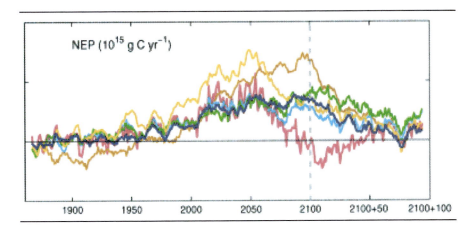

Fig. 3.2.2.3 Net Ecosystem System Production Source: Cramer et al. 2001

Whether terrestrial sinks should actually be used depends on the economic and ecological effects of the alternatives. The IPCC reports mitigation costs through forestry which range from 20 to 100 US-Dollar per ton of carbon in developed countries and only 0.1 to 20 US-Dollar in some tropical developing countries (IPCC 2001c). Depending on the total level of activity these costs vary considerably. However, it is clear mitigation through forestry in developed countries lies within the range of emission control costs of the Kyoto Protocol without hot air and without emission trading (Klepper, Peterson 2003)[2]. For tropical developing countries forestry provides a low cost option for increasing sinks although these countries have so far no commitment for reducing their carbon flows into the atmosphere. These sink options, however, could be effectively used in CDM projects as Figure 3.2.2.4 of the IPCC TAR illustrates which compares the cost curves for different mitigation options. At all mitigation levels forestry in developing countries turns out to be the least cost option.

The research results indicate that the enhancement of terrestrial carbon sinks can be an important part of controlling the atmospheric CO_2-concentration. However, this option seems to be available only within this century as an intermediate supporting measure before the emissions for fossil fuels are drastically reduced. The economic cost estimates also indicate that carbon sequestration could be a cost-effective mitigation option. These rough estimates will need to be analysed with more scrutiny in order to get more reliable results on specific management options on a regionally and sectorally disaggregated level.

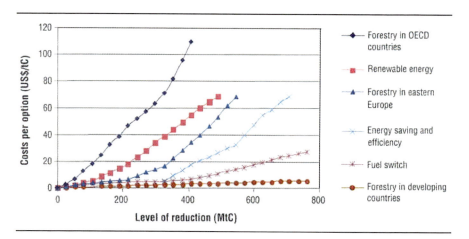

Fig. 3.2.2.4 Cost Curves for Emission Reduction and Carbon Sequestration Source: IPCC (2001c)

---

[2] For the effect of hot air trading see Klepper, Peterson (2002)

The final carbon flux that could potentially be managed is that from the atmosphere into the ocean. Ocean uptake due to the human induced increase in the atmosphere is in the order of 1.9 GtC (see Figure 3.2.2.2). The IPCC Report on mitigation (IPCC 2001c) discusses several options such as ironing of the ocean in order to increase the marine net primary production. The effects of such activities are not clear so far, neither in terms of the amount of carbon that could be removed over and above the natural fluxes, nor in terms of the side effects on marine ecosystems. Even the potential quantities which could be removed from the atmosphere and transported permanently into the deep ocean are not known so far. Finally, the option to capture carbon at the end of the pipe of industrial processes is an option that is technically feasible and apparently not prohibitively costly. It is also unclear as to how much this option can contribute to the removal of carbon from the human carbon cycle.

In summary, given today's knowledge carbon management going beyond emission control is worth wile. The terrestrial carbon storage could be enhanced significantly. Without this sink the effect of the fossil fuel burning for the climate would be much more severe. This option is open for several decades. It is also cost-efficient to do it, as the costs of emission control are much higher than the costs for improving terrestrial carbon storage in some areas. In contrast to terrestrial carbon pools, the amplified use of the ocean store is not sufficiently researched yet in order to assess the potential and the costs of this option. These findings then raise the question whether and how such an extended carbon management can be introduced.

Implementing an International Carbon Management

Changes in the terrestrial carbon stock consist of the net effect of the uptake of carbon in plants and soils and the release of carbon into the atmosphere through land-use change. Both effects cannot be completely separated and both are subject to large natural variations, which are not well understood. However, it is evident which activities might enhance the terrestrial carbon stock. The activities intended to increase the net flow of carbon from the atmosphere on to the earth are local activities mostly in agriculture and forestry. A local or national authority could regulate these activities and thus improve the sink capacity of the area under her control. Unfortunately, the benefit of providing an additional carbon sink is a global positive externality with private costs for achieving it. Consequently – as described above – national authorities have little incentives to provide these sink capacities unless they get reimbursed in some way or the other for their activities. Such compensation can be achieved by an agreement among states or among firms, however, only if some legal background for such private compensation is provided. In both cases, a direct regulation of activities is not sufficient since compensation requires a recognisable service that is provided and, most importantly, is measurable in terms of the quantities of carbon sinks provided.

A prerequisite for an international carbon management with respect to atmosphere-land carbon flows therefore is an accounting procedure for the sink services that a national authority can provide for the international community. Such accounting procedures are subject to a number of difficulties:

There exists considerable uncertainty with respect to the dynamics and the drivers for the development of sinks. It may be particularly difficult to separate natural from human-induced changes in sinks. As credits can only be claimed for human sink activities and not for natural variations (e.g. due to El Niño events), analytical tools for decomposing the effects would be necessary and would need to be accepted by the international community.

Creating terrestrial sinks is a reversible process, i.e. the carbon could be mobilised at later times. Therefore, accounting need not only cover sink activities, it will also need to cover the continued maintenance of established sinks, or their decrease due to a remobilisation of carbon.

The ability to store carbon in terrestrial sinks also depends indirectly on human activities. Historic emissions for GHGs and the deposition of nitrogen both can potentially have positive impact on terrestrial sinks by enhancing the production of biomass. During COP7 in Marrakech these indirect effects were excluded from the definition of human induced uptake of carbon. Again, it will be difficult to separate both.

Reducing emissions for CO_2 is essentially an irreversible process, once they have been avoided in a specific period, the decision cannot be reversed. There is only the option in later periods not to continue avoiding emissions. Terrestrial carbon sinks can be mobilised over time, hence they are reversible, even without human interference. This would mean that emissions avoided should be treated differently from sinks created in an accounting system. This creates difficult legal problems, e.g. who is liable for carbon sinks and for how long such liability should last.

The prerequisite for an internationally agreed joint carbon management that includes all terrestrial sinks is a comprehensive "full carbon accounting". Only such a system could overcome the incentives for free riding by national governments. Since "full carbon accounting" does not seem a feasible option, present international agreements have limited the sink issues to those where accounting can be done with sufficient accuracy. This partial accounting refers to forestry activities (Art. 3.3 Kyoto Protocol) and can be extended in future commitment periods if appropriate measurement concepts are available.

The current climate policy framework consisting mainly of the Kyoto Protocol does not provide for a comprehensive carbon management. First, because as just described the tools for administering potential commitments are not available. Second, because only a fraction of states have committed themselves to reducing carbon flows into the atmosphere and increasing those into the terrestrial pools. This second problem of carbon management has to do with the nature of the international coordination of policies. Participation is voluntary and therefore a state will only join an international institution for carbon management if it can expect a net gain to its state from doing so. However, this is in general not the case since the benefits of emission controls are global, their costs are local. Providing sinks creates incentives for Annex-B countries since they can substitute emission reductions by sink improvements, but this is not the case for Non-Annex-B countries, except possibly for CDM projects.

This architecture of the Kyoto-Protocol with commitments, compliance regimes and punishment for non-compliance – which incidentally is not determined yet – makes it difficult if not impossible to come to a broad agreement. With the USA not signing and other big emittors of CO_2 such as China and India refusing to make any significant commitment for reducing emissions or enhancing sinks this problem has become most evident. Barrett (2002, 2003) therefore advocates a switch from negative incentives – i.e. compliance control and punishment – to positive incentives. These include the help with the development and diffusion of carbon friendly technologies. Such positive incentives increase the participation of states because they surely gain from such a technology transfer. It has a second advantage, they do not require a comprehensive accounting system for emissions. On the downside, they do not necessarily eliminate all climate-damaging activities.

This switch from negative to positive incentives can also be a policy option for the management of terrestrial sinks. The dissemination of knowledge and support for land management practices that can be applied by farmers and forest managers may enable them to lead a successful business and simultaneously provide carbon sinks for the international community. Such activities could be supported by a carbon fund that subsidises carbon friendly land-use practices. In this way not only the incentive problem is reduced or even eliminated, it also solves the measurement and accounting problem. It would no longer be necessary to try to develop a full carbon accounting for every patch of land, a truly Herculean task. Instead, it would suffice to account for the use of the funds and their approximate usefulness in terms of providing carbon sinks.

Such an approach may be quite far away from an efficient carbon management regime. However, since the full regime can not be implemented currently for reasons of limited understanding of processes, for a lack of appropriate accounting, and for a lack of incentive to join an international institution for carbon management, supporting the reduction of carbon flows from land-use change into the atmosphere and the increased capturing of carbon in terrestrial sinks could be a second best alternative.

References

Barrett SA (2002) Towards a better climate treaty. Nota di Lavoro 54, Fondazione Eni Enrico Mattei

Barrett SA (2003) Environment and statecraft – the strategy of environmental treaty making. Oxford University Press, Oxford

Cramer W et al. (2001) Global response of terrestrial ecosystem structure and function to CO_2 and climate change: results from six dynamic global vegetation models. Global Change Biology 7:357-373

Heister J (1997) Der internationale CO_2-Vertrag: Strategien zur Stabilisierung multilateraler Kooperation zwischen souveränen Staaten. Kieler Studie 282. J.C.B. Mohr, Tübingen

Houghton JT et al. (eds) (2001) Climate change 2001: the scientific basis. Working group I contribution to the Third Assessment Report of the IPCC. Cambridge University Press, Cambridge

Klepper G, Peterson S (2003) Trading hot-air. Kiel Working Paper 1133. Kiel Institute for World Economics, Kiel

Klepper G, Peterson S (2003) On the robustness of marginal abatement cost curves: the influence of world energy prices. Kiel Working Paper 1138. Kiel Institute for World Economics, Kiel

McCarthy JJ et al. (eds) IPCC Climate Change 2001 – Impacts, adaptations and vulnerability. Contribution of working group II to the third assessment report of the Intergovernmental Panel on Climate Change. Cambridge University Press, Cambridge

Metz B et al. (eds) IPCC (2001c) Climate Change 2001 – Mitigation. Contribution of working group III to the third assessment report of the Intergovernmental Panel on Climate Change. Cambridge University Press, Cambridge

Noble IR (2001) Ocean and land carbon dynamics: sinks forever versus sink saturation. CRC for Greenhouse Gas Accounting. http://www.greenhouse.crc.org.au

Ott K et al. (2003) Reasoning goals of climate protection – specification of Art. 2 UNFCCC. Research report to the Federal Environmental Protection Agency of Germany (UBA), manuscript

Watson RT, Noble IR (2001) Carbon and the science-policy nexus – the Kyoto challenge. The World Bank, draft

3.3 Food Systems

3.3.1 Global Environmental Change and Food Systems GECAFS[1]: A new interdisciplinary research project

John Ingram[1], Mike Brklacich[2]

[1]GECAFS International Project Office, NERC-Centre for Ecology and Hydrology, Wallingford, OX 10 8BB
[2]GECHS[2] International Project Office, Carleton University, Ottawa

Global environmental change is happening. Human activities, including those related to food, are now recognised to be partly responsible for changing the world's climate and giving rise to other, globally and locally important environmental changes. These include alterations in supplies of freshwater, in the cycling of nitrogen, in biodiversity and in soils.

There is growing concern that the ability to provide food – particularly to more vulnerable sections of society – will be further complicated by global environmental change (GEC).

There is also concern that meeting the rising societal demand for food will lead to further environmental degradation, which will, in many cases, result in further uncertainties for food provision systems.

Policies need to be formulated that enable societies to adapt to the added complication GEC will bring to food provision, while promoting socio-economic development and limiting further environmental degradation. Such policy formulation needs to be built upon an improved understanding of the links between GEC and food provision. "Global Environmental Change and Food Systems" (GECAFS) is designed to meet this need.

---

[1] Global Environmental Change and Food Systems (www.gecafs.org) is a Joint Project of the International Geosphere-Biosphere Programme (IGBP), the International Human Dimensions Programme on Global Environmental Change (IHDP) and the World Climate Research Programme (WCRP).

[2] Global Environmental Change and Human Security Project (www.gechs.org) is a Core Project of IHDP.

Background

Food is fundamental to human well-being. Improved methods are needed to grow, harvest, store, process and distribute food as societal demand for agricultural and fisheries products increases, and in many parts of the world economic and social development is often mediated by food constraints at local and regional levels.

Links between food systems and the environment are well documented. Environmental factors such as climate, soils and water availability have long been recognised as major determinants of the ability to produce food in a given location, and a wide range of farming and fishing strategies have been developed in response to the differing environmental conditions around the world. The production, processing and distribution of food however have considerable impacts on environment by, for instance, altering biodiversity, emitting greenhouse gases, and degrading soils and other natural resources by overexploitation and pollution. This close, two-way relationship with environment exerts considerable influence on production and – ultimately – on the availability of, and accessibility to food.

Until recently, the effects that food systems have on environment were perceived at relatively local spatial scales. For example, soil erosion caused by intensive crop production resulted in the siltation of nearby water courses, and contamination of ground and surface water supplies by agricultural chemicals did not reach beyond local water sources. However, human activities – in considerable part due to satisfying the need for food – are now recognised to be changing the environment over large regions, and even at global level. Overall these macro-scale changes can be divided into two broad categories. One involves fundamental changes to major earth systems and functions which operate at the global level, such as climate and the cycling of nitrogen. The other involves incidences of environmental change at the local level which are so widespread as to be considered global phenomena; degradation of fresh water resources and soil erosion have, through their collective extent, transformed from local concerns and are now issues that must be considered and addressed over large regions. Collectively these changes are termed "Global Environmental Change" (GEC). GEC will bring additional complications to the already difficult task of providing sufficient food of the right quantity and quality to many sections of society. Improving food provision in the face of GEC, while at the same time minimising further environmental change, is a crucial issue for both development and society at large.

Food Provision and Food Systems

Recent years have seen a greatly increased understanding of how GEC will affect food productivity at field level, and research results pave the way for broader analyses of GEC impacts on food production on a regional basis. However, there is a

---

Box 3.3.1.1 Food Provision

Food provision is governed not only by production, but also by the availability of, and access to food. Access to food is a function of economic potential, physiological potential (e.g. nutritional quality) and food availability (which depends on production and distribution). Food production is a function of yield per unit area and the area from which harvest is taken.

Production = f (yield, area)
Availability = f (production, distribution)
Access = f (availability, economic & physiological potential)

Food Provision = f (Production, Availability, Access)

---

need to think beyond productivity and production – of ultimate interest is food provision, a concept of greater relevance to society well-being and hence policy making.

A wide range of sciences is needed to address the components of the "Food Provision Equation": estimates of food production are founded in agroecology, agriculture and fisheries sciences, while issues related to distribution are largely researched by social and policy-related sciences. The broader notion of access requires consideration of a further set of disciplines including economics, sociology and nutritional sciences.

Akin to the need for adopting the broader concept of food provision (rather than just food production), research planning and policy formulation needs to be set within the context of *food systems*, rather than just food supply. Developing research in the context of food systems helps to identify and integrate the links between a number of factors "from plough to plate" (Atkins and Bowler 2001), including consideration of production, harvesting, storage, processing, distribution and consumption. The approach thereby allows a more thorough understanding to be developed of the "impacts" and "feedbacks" links between food provision and environment. It will also help to identify where technical and policy interventions might be most effective to (i) cope with short-term impacts of GEC; and (ii) help adapt for environmental conditions in the longer term. Coping and adaptation strategies for food provision will however need to differ depending on the degree to which people and communities are vulnerable to the impacts of GEC. Not all individuals and sections of society are equally vulnerable to GEC; their capacity to cope with existing variability in biophysical and socio-economic systems, and their ability to perceive GEC and adapt food systems accordingly vary considerably. This is because these factors are controlled by the flexibility with which the supply, availability and access to food (and related, essential resources) is mediated by socio-economic institutions such as land tenure, access to credit, exploitation rights of renewable resources, etc.

Adaptation strategies will need to encompass both biophysical and policy options. Management decisions must however be underpinned by a sound understanding of both the socio-economic and environmental consequences that different possible strategies will bring.

GECAFS: A New Research Approach

The interactions between global environmental change and food provision involve many complex issues spanning natural, social and climate sciences. The International Geosphere-Biosphere Programme (IGBP), the International Human Dimensions Programme on Global Environmental Change (IHDP) and the World Climate Research Programme (WCRP) already encompass broad research agendas in these three major areas. However, in order to advance our understanding of the links between GEC and food provision (and thereby help to develop and promote effective interventions) IGBP, IHDP and WCRP have launched the Global Environmental Change and Food Systems (GECAFS) Joint Project as a new, interdisciplinary approach. Furthermore, the research agenda is broader than impact studies alone (important though these continue to be) as it explicitly includes research on how food provision systems could be adapted to the additional impacts of GEC, and the consequences of different adaptation strategies for socio-economic conditions and environment. By including both "impacts" *and* "feedbacks" in the context of food provision a niche for new research is clearly defined.

GECAFS has been conceived to address issues of interest to development, to society at large, as well as to science. An innovative, three-way dialogue between policy makers, donors and scientists is being established to develop specific research agendas which are useful to aid policy formulation, scientifically exciting and fundable. Many research groups are active in the general area of food "security" but their activities generally focus on *current* impediments to food production. Building on ongoing studies but emphasising GEC issues, and linking closely to the needs of policy formulation, the structured approach will deliver an efficient research mechanism to address the rapidly emerging "GEC-Food" agenda. Of ultimate interest is the link between GEC and societal well-being (rather than with food systems *per se*). This however has to be addressed through the researchable issues needed to understand the relationships between GEC and food systems and it is this that requires the innovative, interdisciplinary approach.

GECAFS Goal and Science agenda

The GECAFS goal is to determine strategies to cope with the impacts of Global Environmental Change on food provision systems and to analyse the environmental and socio-economic consequences of adaptation.

Research is being developed as three, inter-related science themes (see Figure 3.3.1.1).

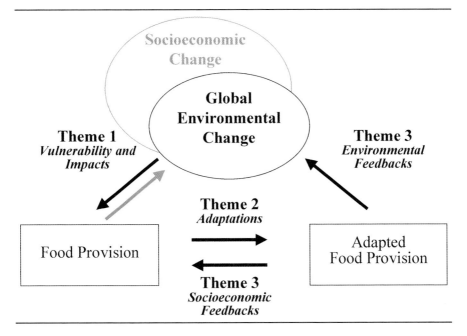

Fig. 3.3.1.1 A diagrammatic representation of the three GECAFS Science Themes with respect to GEC and food provision systems. The contextual issues of changing socio-economic conditions and the consequences of current food provision systems on GEC are depicted in grey, while the main features of GECAFS are shown in black

- Theme 1 : Vulnerability and Impacts: Effects of GEC on Food Provision
- Theme 2 : Adaptations: GEC and Options for Enhancing Food Provision
- Theme 3 : Feedbacks: Environmental and Socio-economic Consequences of Adapting Food Systems to GEC.

Theme 1 "Vulnerability and Impacts"

Theme 1 research is set within the context of the question "Given changing demands for food, how will GEC additionally affect food provision and vulnerability in different regions and among different social groups?". This question recognises that many factors already affect food provision and vulnerability, and that these are posing different stresses. It however raises the issue that GEC may well bring further complications – hence the word "additionally" – and further recognises that vulnerability to GEC varies for different food provision systems, and hence will have differing impacts among different social groups.

Food provision is controlled by a range of biophysical and socio-economic factors working interactively at a range of temporal and spatial levels. These factors ultimately determine the vulnerability of food systems to both biophysical and socio-economic

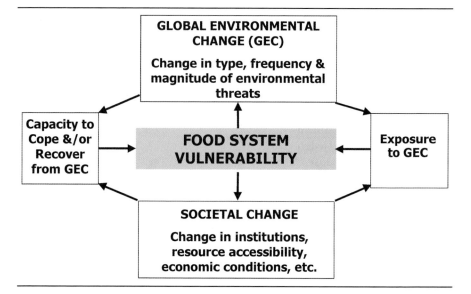

Fig. 3.3.1.2 Human vulnerability *vis à vis* global environmental and societal changes. (Figure derived from Bohle 2001, McMichael and Githeko 2001.)

change (see Figure 3.3.1.2). Biophysical factors include climate, weather and site-related natural resources (e.g. soils, topography, water availability, previous vegetation and site management, distribution of exploited fish populations, coral reefs); socio-economic factors include current agricultural, aquaculture and fisheries management (e.g. germplasm selection, timing of operations, nutrient and pest management), population density and demand for food products (for local consumption and trade), availability (markets, distribution, storage) and access (e.g. socio-political controls, exploitation rights, equity, wealth). Research will therefore address constraints and opportunities for meeting future demands for food from several perspectives including aggregate regional supply and demand, and broad scale socio-economic conditions which either threaten or promote food accessibility. Theme 1 will identify where GEC will be particularly important and why, and also examine the crucial issue of vulnerabilities and impacts of GEC on regional food production potential.

Theme 2 "Adaptations"

Theme 2 deals with the basic question "How might different societies and different categories of producers adapt their food systems to cope with GEC against the background of changing demand?". This question recognises that within given societies not all people and groups will be equally able to adapt to changing demands for food, and that adaptation strategies to cope with the additional complication of GEC will vary; different groups will have different limitations to adaptation.

Theme 2 aims to understand how communities and institutions might anticipate, resist or adapt to, and recover from the impacts of GEC. GEC may cause
food provision problems in the short term, when adaptation mechanisms are
too slow, and in the long term, where adaptation mechanisms are simply not
effective. Research needs to concentrate on how environmental and socio-economic forces combine at local to regional levels either to enhance or reduce vulnerability; on existing strategies for coping with food shortages; and on the extent
to which global environmental and socio-economic forces might alter human vulnerability within selected regions. To make these assessments will require the
identification of the most vulnerable groups, the key institutions in society that
make or break coping and adaptation efforts, and management efforts which will
be needed to counteract the negative aspects of GEC. The nature of critical
thresholds for adaptation, and when and where these will be exceeded will be a
critical part of the research agenda. In addition, it will determine the speed of
coping and adaptation that different groups in society possess, and how this
relates to GEC.

Theme 3 "Feedbacks"

The third theme considers "What would be the environmental and socio-economic
consequences of alternative adaptation strategies?" The question recognises that different adaptation strategies will have different consequences for both socio-economic systems and environment, and that both types of consequence are equally
important and need to be considered simultaneously. Theme 3 will allow regional
level analyses of "tradeoffs" between socio-economic and environmental issues for
a range of management and policy options. These analyses will be conducted within
the scenarios agreed upon in the project planning phase.

Research will need to develop tools to identify, and quantify as far as possible,
the feedbacks to environmental issues such as atmospheric composition and other
climate change drivers, consequences for biodiversity and land and aquatic degradation, and also to socio-economic issues such as livelihoods, institutional flexibility and policy reform. The rapidly growing concern for potential environmental
degradation due to changes in genetic variability and biotechnology also needs consideration. By complementing work in Themes 1 and 2, this "feedbacks" component will identify GECAFS as a comprehensive GEC research programme.

The GECAFS science themes provide an innovative framework within which
new areas of science can be developed and harnessed to address societal concerns.
Examples include:

- Methods for the analysis of environmental and socio-economic tradeoffs in food
  systems.
- Analyses of changing human wealth and food preferences and interactions with
  biophysical models of GEC to produce new insights of regions where food provision may be sensitive to GEC.

- Methods to allow the appropriate level of aggregation of small scale food production systems and disaggregation of global scale scenarios and datasets to address regional and subregional issues.
- Comprehensive scenarios of future socio-economic and environmental conditions involving the food system; socio-economic and demographic factors; and environmental and ecological data.
- Analyses and insights into the combined institutional and technological factors which can reduce societal vulnerability to GEC.
- Developing combined socio-economic-biophysical indices of vulnerability.
- Use of past records of social adaptations to biophysical changes to provide inputs to scenario-based models of the future.

GECAFS regional studies will however need to be set within clearly defined sets – or "scenarios" – of future biophysical and socio-economic conditions. These will be specifically designed to assist analysis of possible policy and biophysical interventions using the interdisciplinary science at the project's core, and will set the context for the individual research projects. They will help to "tease out" the meaning of "socio-economic change" in the context of GEC. The development of these comprehensives scenarios is in itself a major research exercise. This has been initiated by defining three broad categories of attributes as a minimum set of required contributing data: the food system; socio-economic and demographic factors; and environmental and ecological data. The further development of "aggregated indicators" to assess vulnerability will be a significant new science output.

Research Design, Implementation

GECAFS aims to help strengthen policy formulation for reducing vulnerability to global change at national to sub-continental scales; and to provide tools and analyses to undertake assessments of trade-offs between food provision and environment in the context of global change. To be effective research will be developed that meets the needs of national and regional policy makers, the principal "clients" for GECAFS research. GECAFS will therefore engage with policy makers early in the research planning process to develop research that directly addresses their needs and maintain close links throughout the implementation and reporting phases. In collaboration with research partners and collaborators, donors and end-users, GECAFS will (i) identify interdisciplinary research topics of mutual interest to science, development and policy formulation; (ii) help in developing databases and future scenarios to explore tradeoffs; (iii) help in the dissemination of results and obtaining feedback from end-users; and (iv) assist in capacity building as part of its research approach.

GECAFS research will thus be implemented in two major ways: (i) *Regional GECAFS projects* at sub-continental-level, "tailor-made" to address particular interests of policy makers, donors and science community; and (ii) *Integrative GECAFS studies* at multi-region to global level, which integrate individual studies. Individual studies will undertake research to address all three science themes.

Both approaches will be underpinned by the development of "vulnerability science" and "scenarios" as discussed above.

GEC and the Food System of the Indo-Gangetic Plain: An Example GECAFS Research Project

One of the initial regional GECAFS research projects concerns the food system of the Indo-Gangetic Plain (IGP). This is largely dependent on rice and wheat grown in rotation and there is growing concern that the productivity of the system is declining, especially the rice component: an assessment of 11 long-term rice-wheat experiments (ranging from 7-25 years in duration) from the region indicates a marked yield decline of up to 500 kg/ha/yr in rice in nine of the experiments (Duxbury et al. 2000). Continuation of these trends will have serious implications for food provision, local livelihoods and the regional economy. As a given season's weather is a major determinant of yield (due to both the direct effects on crop growth and indirect effects related to management), there is concern that changes in climate, especially related to changes in climate variability, will exacerbate the observed trend. Moreover, other analyses (e.g. Grace et al. 2001) show that the highly intensive production approach currently practiced in large parts of the region is a major source of greenhouse gases, while the current irrigation practice is having serious negative effects on local water tables and water quality.

In the face of GEC, policy requirements are to develop strategies that promote:

- agricultural competitiveness while limiting further environmental degradation;
- food provision systems which enhance the social security of the more vulnerable; and
- rural employment opportunities thereby reducing intra-IGP labour migration and urbanisation.

Due however to the marked socio-economic and biophysical differences across the region, a single approach is not appropriate. A consultation process with local and regional policy makers determined information needs in relation to GEC for regional policy formulation, and gave rise to a number of possible research issues. The process also helped to characterise the regional differences between western and eastern and sections of the IGP:

The western region of the IGP is characterised by high investment in infrastructure, institutions and effective policy support; intensive agriculture; high use of agrochemicals and groundwater for irrigation; high productivity (and hence a food surplus region so making it highly significant for national food security); and seasonal in-migration of male labour (which can lead to social conflicts).

Interdisciplinary research will be developed to address questions such as:

Theme 1

- How will GEC (especially climatic variability) and increasing non-farm demands affect change in water supply and demand and consequent food system vulnerability?

Theme 2:

- How can changes in water management (e.g. through enhanced policy instruments, land use strategies and community participation) and energy-efficient technologies reduce vulnerability of food systems to climate variability and other aspects of GEC?
- How can increasing urban and agricultural wastes and water of inferior quantity be utilised in agriculture to adapt to reduced land and water availability?
- Where, what forms and how much additional public and private investment would be needed to increase on-farm income, maintain water balance and diversify from rice-wheat system?
- How can policies and institutional arrangements best be adapted to promote adoption of existing technology options to enhance production in the face of GEC while conserving natural resources?

Theme 3:

- What will be the consequences of alternative approaches to water management and resource-conservation strategies on rural livelihoods, intra-regional trade, carbon sequestration and GHG emissions, and water tables?
- What would be the consequences of diversifying from rice-wheat system on food production, government procurement, energy use, income and employment potential, resource conservation and GHG emissions?

In contrast the eastern region of the IGP is characterised by a high risk of both flooding and drought; poor infrastructure and limited capacity for private investment; largely subsistence agriculture; low productivity, and hence a food deficit region (leading to poverty, hunger and malnutrition); and out-migration of male labour (leading to increased involvement of women and children in the agricultural sector).

Interdisciplinary research will be developed to address questions such as:

Theme 1:

- How will GEC affect vulnerability of resource-poor farmers to flooding and drought, and how will this exacerbate existing socio-economic inequities?

- What early warning systems of environmental change and its potential impacts would assist stakeholders to identify regions and communities of potentially greater insecurity?

Theme 2:

- What investment policies (e.g. insurance) would encourage farmers and society to adopt available technological options to reduce vulnerability to GEC?
- What infrastructure, market opportunities and technical options need to be developed for diversifying crops (e.g. to aquaculture) to make more effective use of flood and groundwater, and what are the social constraints (e.g. food preferences) to their adoption?
- What policy interventions are needed to reduce the number of hungry and/or undernourished people (especially women and children) considering that food production systems may become even more risk-prone?
- What new institutional mechanisms in research and extension (e.g. involvement of NGOs and private sector) would facilitate generating, adapting, disseminating and utilising knowledge in managing increased risks due to GEC?

Theme 3:

- How would diversification and increased government interventions (e.g. markets, roads, credit, flood control and extension services) affect food provision, rural income, equity, labour migration and employment, water use and quality, biodiversity, and GHG emissions?

A more detailed research planning exercise for the IGP is now being initiated with national policy makers and research groups; with international collaborators including the CGIAR, FAO and WMO; and with scientists from IGBP, IHDP and WCRP. GECAFS will add value to the individual efforts of all its research partners by building on their complementary skills and contributions; it will not "replace" their existing efforts, but draw upon them, and set them in a broader canvas of societal concern.

References

Atkins P, Bowler I (2001) Food in society. Arnold, London

Bohle HG (2001) Vulnerability and criticality: perspectives from social geography. Update, IHDP 2/01:1-5

Duxbury JM et al. (2000) Analysis of long-term soil fertility experiments with rice-wheat rotations in South Asia. In: Abrol IP et al. (eds) Long-term soil fertility experiments in rice-wheat cropping systems. Rice-Wheat Consortium for the Indo-Gangetic Plains, New Delhi, pp vii-xxii

Grace PR, Jain MC, Harrington LW (2001) Global environmental impacts from conservation agriculture. In: Proceedings of the international workshop on "Conservation Agriculture for Food Security and Environment Protection in Rice-Wheat Cropping Systems", February 6-9, Lahore, Pakistan

McMichael A, Githeko A (2001) Human Health. In: McCarthy J et al. (eds) Climate Change: Impacts, Adaptation, and Vulnerability. working group II contribution to the Third Assessment Report of the IPCC. Cambridge University Press, Cambridge, pp 451-485

3.3.2   Food Security in South Asia

Wolfgang-Peter Zingel

South Asia Institute, University of Heidelberg, Department of International Economics
Im Neuenheimer Feld 330, 69120 Heidelberg

The fact that food security is less a question of harvests and quantities rather than of entitlement and deprivation has convincingly been proved by Amartya Sen, the Indian Nobel prize winner in economics of 1998 in his seminal work on poverty and famine (Sen 1994). The distinction, however, was not awarded for his contributions to solving one of mankind's most elementary problems, but for his earlier more academic work. He had spent his youth in Dhaka, now Bangladesh, during the Bengal famine of 1943, which is said to have cost three million lives, but not so much because of a lack of food but because of a short sighted policy, lack of insight and mismanagement (Dréze et al. 1995, Dréze 1999, Knight 1954, Ravallion 1987). Sen's study of this and other famines of the twentieth century shows that problems of food security always need detailed analysis before any solutions may be recommended.

South Asia, where one or two famines occurred almost every decade in the nineteenth century, today feeds its population much better than in the past (Jalan 1997:136, Blyn 1966) although their number has more than triplicated since Independence (India and Pakistan: 1947). But this is only on average. There are still more people undernourished in South Asia than in any other world region despite the fact that the Government of India is sitting on unprecedented large stocks of food grain as well as foreign exchange reserves.

The following text will focus on the macro rather than the micro aspects of the problem. It will start with asking some basic questions: What kind of food has to be secured and for whom, why and how? Where should we draw the dividing line between food security and insecurity? And looking at the economics of food security: Has food to be produced locally? At what prize may food security come? And finally, who is to pay? The first question, however, is why we should look at South Asia.

Why South Asia?

South Asia, or India, as the region south of the Himalayas was known until Independence, is home to one fifth of mankind. Together with Africa south of the Sahara South Asia stands for food problems, for natural calamities like floods and droughts. But South Asia has seen remarkable improvements; thanks to the "green revolution" the overall food position is better than at any time in recent history (Chopra 1986). But only on average of the population. South Asia still sees some of the worst forms of abject poverty and lack of food (Bohle 1997, Zingel 1999). There is no standard definition of South Asia. For all practical purposes the seven states that constitute the South Asian Association of Regional Cooperation (SAARC -SA7), i.e. India, Pakistan, Bangladesh, Sri Lanka, Nepal, Bhutan and the Maldives, belong to South Asia. They all were ruled directly or indirectly and for longer or shorter times by the British colonialists. The same applies for Myanmar (then Burma) and Afghanistan (much forgotten now). But Myanmar joined the Association of South East Asian Nations (ASEAN) and Afghanistan the Economic Cooperation Organisation (ECO - but so did Pakistan), too. They see themselves as parts of South East Asia and Central Asia, respectively, rather than of South Asia. In any case, the two countries have only less than five percent of South Asia's (SA9) population; including them or not in our analysis would not yield very different results. After Independence the countries of the region maintained much of their colonial administrative system. Only India and Sri Lanka succeeded in maintaining (more or less) a democratic political system with political parties, fair and free elections based on universal adult franchise. All states pursued a highly interventionist if not socialist economic policy, especially with regard to food security. South Asia, thus, presents an interesting example for a case study on (national) food security.

What kind of food?

South Asians are known to be vegetarians, eating rice as staple food plus some pulses and vegetables. This is true, of course, but needs differentiation, as the Anthropological Survey of India found out (Singh 1992). Up to one half of South Asia's inhabitants may be counted as upper caste ("non-scheduled") Hindus, who traditionally – although not necessarily – are vegetarians; this also applies to Buddhists, but only in principle (as can be experienced in Bhutan and Myanmar). Low ("scheduled") caste Hindus, most of the tribal (often Hindu) population, all of the 400 million Muslims (constituting the largest Muslim population in any world region) and the Christians are more likely to be non-vegetarians, with the major exception of pork in the case of the Muslims. If meat is eaten, it is mainly for the sake of taste, i.e. in small quantities. Not eating meat may not suffice in India to make one a vegetarian: eggs, for example, are usually considered to be non-veg food; vegetarians may also shun dairy products and even vegetables like onions and garlic. On the other hand, chicken meat used to be allowed on meatless days. But usually vegetables, especially onions, are essential complements, irrespective of rice or wheat (in the north west) being the main source of food energy. Vegetarian in the South Asian context does not mean raw vegetable as is often the case in the West: almost all food is cooked and fuel (mostly shrub, sticks, dung, wood), thus, is as important for feeding people as all the various food items. This is even more true for (drinking) water, which as a beverage is the most important food item and is also needed for almost all food preparations.

Food items can be substituted by other food items only to some extent, not the least for practical purposes: one needs different cooking utensils (and sometimes even different ovens and fuel) for example for cooking rice or baking the various kinds of "bread". Adaption to new food items usually is slow – and irreversible. Food is also more than food energy; but if the energy intake is insufficient, proteins may be burnt as food energy, which increases malnourishment: deficiencies in protein and fat are more pronounced than deficiencies in energy intake.

Food security

Food, although seemingly available in sufficient quantities on aggregate levels (e.g. the region or a country) and over longer periods, may nevertheless be available only scantily for certain areas, groups, individuals and for shorter periods of time, and even that on an irregular basis. For many even one meal per day on all days of the year would be a remarkable improvement. If this affects a substantial part of the population over longer periods we would recognise this as the typical manifestation of a famine. But for most of those suffering from food insecurity the typical outcome would be less than average weight (wasting) and short size (stunting) and a greater probability to succumb to any illnesses. In short: statistically, food security becomes a matter of the level of aggregation and of probabilities.

Food security for whom?

We can observe groups with more pronounced deficiencies of (a) total food, (b) essential food items and food complements, (c) coping capacity (for lack of savings) and, thus, (d) a higher risk (or higher vulnerability). On the national level, Bangladesh and Nepal (and Afghanistan) lack food security more than the others, as can be seen from Table 3.3.2.1 (cf. also: FAO). Inside India, the so called Hindi or cow belt of northern and central India plus Orissa suffer more from insecurity than the rest of the country. In general, the "weaker sections" or poorer ones can look forward to less food security, they may be urban or rural population, often, although not necessarily, lower caste or tribal. There is a substantial gender bias when it comes to in food; children in South Asia are less well fed than those in Africa. Accordingly, girls from low caste families in backward areas are among the worst fed (Smith and Haddad 2000).

Table 3.3.2.1 South Asia. Food availability per head and day

Country/ Years	Calories (Kcal)		Proteins (g)		Fats (g)	
	Total	Animal Product	Total	Animal Product	Total	Animal Product
Bangladesh						
1961-1963	2,090	63	42.9	5.4	15.5	3.8
1969-1971	2,122	67	45.2	6.1	15.3	3.9
1979-1981	1,975	60	43.6	4.9	14.6	3.6
1989-1991	2,065	60	44.4	4.9	17.8	3.5
1998-2000	2,101	63	45.1	6.1	21.5	3.9
Bhutan						
1966-1968	2,050	35	45.1	1.7	22.1	2.6
1969-1971	2,065	35	45.5	1.7	22.2	2.6
1975-1977	2,058	34	45.4	1.7	22.0	2.6
India						
1961-1963	2,048	112	52.5	6.1	31.4	7.5
1969-1971	2,041	105	51.0	6.0	30.3	7.0
1979-1981	2,083	120	50.8	6.8	33.3	7.9
1989-1991	2,365	162	57.4	9.0	41.2	10.9
1998-2000	2,426	194	57.8	10.5	47.2	13.0
Maldives						
1961-1963	1,545	151	45.3	22.4	32.0	6.1
1969-1971	1,624	192	53.2	28.5	37.1	7.7
1979-1981	2,165	185	69.2	28.1	47.0	14.2
1989-1991	2,365	305	72.0	28.3	47.0	14.2
1998-2000	2,578	650	112.9	79.3	65.1	30.0
Nepal						
1961-1963	1,833	148	47.7	7.7	26.8	10.2
1969-1971	1,832	150	48.5	7.9	25.2	10.5
1979-1981	1,891	158	50.3	8.6	26.3	11.0
1989-1991	2,443	159	63.0	9.0	32.3	11.1
1998-2000	2,381	160	61.0	9.2	34.4	11.2
Pakistan						
1961-1963	1,786	273	49.7	13.8	31.4	18.7
1969-1971	2,223	274	56.1	14.0	35.6	18.7
1979-1981	2,173	273	52.0	14.4	44.9	18.5
1989-1991	2,323	336	57.9	17.8	55.5	23.2
1998-2000	2,458	437	62.0	22.6	54.7	29.8
Sri Lanka						
1961-1963	2,138	105	43.6	9.0	46.0	5.9
1969-1971	2,289	104	46.5	8.9	46.2	5.9
1979-1981	2,348	114	46.8	9.4	46.9	6.5

Table 3.3.2.1 South Asia. Food availability per head and day—cont'd

Country/ Years	Calories (Kcal)		Proteins (g)		Fats (g)	
	Total	Animal Product	Total	Animal Product	Total	Animal Product
1989-1991	2,222	124	47.6	10.4	43.9	6.9
1998-2000	2,360	156	53.4	13.7	44.8	8.5
Afghanistan						
1966-1968	2,170	179	67.5	10.3	29.3	12.4
1969-1971	1,987	158	61.6	9.1	21.6	11.1
1975-1977	1,974	147	60.8	8.4	27.0	10.1
1994-1996	1,706	146	47.1	8.5	26.2	11.0
Myanmar						
1961-1963	1,770	93	45.7	8.2	30.4	5.1
1969-1971	2,040	96	52.6	7.7	33.0	6.9
1979-1981	2,327	108	60.1	8.5	35.5	7.2
1989-1991	2,620	98	65.0	8.2	41.7	6.4
1998-2000	2,823	120	72.6	9.7	45.1	7.8

Notes: Three years averages. No up-to-date information on Afghanistan and Bhutan. *Sources*: FAOSTAT, 10 Dec 2002, except: Afghanistan and Bhutan 1966-68, 1969-71, 1975-77: FAO production yearbook 33.1979, pp. 61-71 und 249-259. – Afghanistan 1994-1996: FAOSTAT, 15 Apr 1999.

Why food security?

The fact, that children in South Asia are less well fed than those in Africa, even at comparable family income levels, shows, that "development", if measured in per capita income, does not automatically bring out the levels of nutrition that could be expected. If we regard the well-being of the people (however defined) established only if a certain level of "basic needs" guaranteed for all the people and none of them suffering from hunger over any longer period of time, then food (supply) for all has to be secured.

How to secure food supply?

Immediately after World War II broke out, the British-Indian government started their public food distribution system (PFDS) which has survived in India and other parts of the erstwhile Empire till today. Burma was the main "surplus" province of British India, the main harbour towns Calcutta and Bombay were the main "deficit" areas, and transport was easy by ship. Once the war started and the Japanese had taken Burma, the supply stalled. Imports from other countries were impossible because of submarines and the navy's own requirements of vessels. Poor weather conditions led to a shortfall in the grain harvest of 1943 and prices skyrocketed because traders withheld ("hoarded") supply in expectation of further price rises. When in this situation the administration of "surplus" provinces decided to close their borders, the poorer sections of Bengal were simply out-priced from the market. The problem was aggravated by the fact that two centuries of colonial rule and a fateful system of heavy land taxes and eviction in case of non-payment of taxes and rents to an ever growing number of absent landlords and middlemen had left so many of the rural population landless and with no other income than from casual labour. Labour was abundantly available and, even at distress levels of wages, not absorbed by the market.

The food administration was unable to solve the crisis, especially not in Bengal where it developed into the worst catastrophe since the eighteenth century. Over the years, and especially outside Bengal, food management became more efficient and when another bad harvest struck in 1946 it could be well managed. Collective memory remembers the famine mainly as the result of the greediness of traders and hoarders; the public distribution system which evolved from the war administration had their ups and downs but has been quite popular especially in India and Bangladesh (De Vylder 1982, Tyagi and Vyas 1990, Zingel 2003).

The other major experience was the poor harvest of 1965, when famine could be averted in India and Pakistan thanks to the food aid from the USA and Canada at unprecedented levels. This aid, however, failed to make an impact politically, because it was tied to the superpowers' (US and Soviet Union) demand to end the border war between the two South Asian countries of the same year. The American vice-president's talk of "food power" made it abundantly clear that the South Asian nations were much less independent than their talk of self sufficiency and self reliance would have people think to be.

India and Pakistan were saved from a further dependency on food aid after the almost instant success of what became known as the "green revolution". Research of the Rockefeller Foundation (i.e. US!) funded Centro International de Mejoramiento de Maiz y Trigo (CIMMYT – International Maize and Wheat Improvement Center) in Mexico had led to the development of high yielding varieties of wheat; seed was available in sufficient quantities when farmers and governments were ready to adopt them under the impact of failure of traditional seeds and production technologies. The "miracle wheat" became an instant success, which, however, was less predicted and planned as governments later used to claim. The new dwarf varieties (because of their short stems) were more demanding as far as quantities and timing of water, fertiliser and pesticide doses were concerned. Fortunately, as another unplanned result of problem management, India and Pakistan had just undergone major irrigation investments, so that water became available when needed.

Partition of British India cut through the system of the Indus and its five major tributaries in the Punjab (= five waters). These rivers receive most of their waters in the high mountains of the Himalayas and the Karakoram, and Pakistan, situated at the tail ends of the rivers, feared to be cut off by India which had started diverting water to its side. A war over water between the two neighbours could be averted only with the help (and money) of a group of friendly western nations. In 1960 the Indus Water Treaty was signed which allotted all of the water of the three western rivers (Indus, Jhelum, Chenab) with around three quarters of all water to Pakistan and all the water of the three eastern rivers (Ravi, Beas and Sutlej) with the remaining quarter of all water to India. Both countries built large dams and link canals: in Pakistan to divert water from the Indus into the Jhelum and on to the Chenab, the Ravi and the Sutlej; India diverted the waters from the Ravi, Beas and Sutlej before they enter Pakistan to the (East) Punjab and to Rajasthan. This solution may be regarded as uneconomical, but turned out to be highly beneficial politically, because the solution is very easily manageable as compared to an end-less bickering over the waters as we can observe in the case of the Ganges, which is shared between India and Bangladesh, but in reality leaves Bangladesh high and dry, as they claim, during the essential pre-monsoon time. Likewise, domestically, the Indian states of Karnataka (upstream) and Tamil Nadu (downstream) have been locking horns over the water of the Cauvary.

The first years of the "green revolution", i.e. the late 1960s and early 1970s, wit-nessed such an increase in wheat production, that India saw themselves becoming a major food grains exporter. But soon India had to experience a (minor) setback in wheat production and the social repercussions of the "green revolution" became evident. In Punjab, landlords (not all of them "big") had started to evict tenants and to cultivate their farms themselves and/or with hired hands. Higher incomes allowed them to buy machinery, again with a labour displacing effect. Ever since the "green revolution" has been brandished for their negative social impact rather for its economic success. But looking at the development of yields, we can clearly make out, that they have been rising ever since and have reached, on an average around thrice the level of the 1950s. The other complaint was that the "green rev-olution" was restricted to wheat and, thus, to the north west of the subcontinent. But rice yields have also improved: not exactly in such a dramatic way as wheat yields, but they more than doubled over the last half century. This benefited the east and the south of the subcontinent. Sorghum and millet, however, which are grown on the high plains of the Dekhan, did not see any improvement in yield and have been grown on fewer areas at marginal locations.

Another reason for the success of wheat and rice was the state's price policy (De Janvry and Subbaro 1986). De-linking their countries from the world market (India more than Pakistan) allowed governments to make their procurements of basic food items at below world market prices. They even continued the (war time) regime of regional price differentiation and restricted movements of food items between the different regions of their countries. They were obviously not aware of the farmers' price elastic reactions, i.e. of the depressing effect of low prices on pro-duction. Government intervention into food markets (Lele 1971) culminated in the early 1970s when Indira Gandhi "nationalised" food grain marketing in India and Zulfikar Ali Bhutto rice and oil mills in Pakistan. Both strategies utterly failed and had to be given up after a few months. Both actions were also fateful for the two top politicians' political careers. Since then the two countries have followed different paths. Pakistan by and by gave up the heavy intervention in the food markets; public food distribution became less important and was finally given up in the 1980s. Sugar was the last item rationed. The system was said to be more and more inefficient and few people relied on it at the end.

India still runs its system of public food distribution. The high food grain reserves mentioned above are, however, more the outcome of a politically motivated price policy. By the end of the 1970s a militant nationalism had evolved in the Indian state of Punjab, fuelled by the central government's (again: Indira Gandhi) attempt to install a regional government of their choice. When the central government finally tried to re-take control of the state (and the Sikhs' Golden Temple in Amritsar), tension had reached dimensions of a civil war; in 1984 prime minister Indira Gandhi was shot by her Punjabi body-guard. Subsequent governments managed to de-escalate the conflict, among others by granting generous support prices for wheat and rice, with Punjab, India's major "surplus" state, benefiting most from the high prices. The public distribution system is still working, but becoming more and more costly – and some say: less efficient (Chopra 1981, Chopra 1988).

Let us analyse, thus, the major instruments of food security in South Asia, on the macro as well as on the micro level.

Instruments for food security I:
Temporal adjustments

Maybe the oldest instrument to secure food supply is to retain a part of the harvest and store it for the future. This always has been done on an individual basis and – as a sign of good governance since times immemorial – on a collective level as a prime responsibility of the state. In Moenja Daro, a centre of the Indus civilisation, structures of a large granary, more than 3,500 years old, can still be found. Since food items lose their quality over time, stocks have to be replenished by newer ones from time to time, a demanding organisational task which requires tight technical and financial scrutiny. Food grains always have been the bulk of food stocks. Inadequate storing facilities and bad management have always been major problems of South Asian food administrations (Chand 2002, 2003). In India the Food Corporation of India is entrusted with most of government storage; there has been an impressive programme to build storage, but still much of the food grain is stored in the open (covered and plinth). Indian authors claim, that only a small percentage is lost due to improper storage. More may be siphoned off as "system losses" or is already unfit for consumption when purchased by corrupt procurement personnel. The enormous amount of present food grain stocks of over sixty million tons in India is an indication that storing no longer is an impossible task.

Instruments for food security II:
Regional adjustments

Until the colonial power built the railways, the "steel frame" of India, famines in one part of the subcontinent could happen while plenty of food was available in other parts; the worst (known) of such famines was one in Bengal in the 1760s, shortly after the East India Company had taken over the *diwani* (government) of the *subah* (province) from the mogul emperor. By the late nineteenth century the railway network allowed the transport of large quantities from one end of the "jewel in the crown" to another. Famines seemed to be a thing of the past. One of the reasons why the Bengal famine of 1943 could become so devastating was, that rice from Burma no longer was available because of the Japanese occupation and food from elsewhere could not be shipped to India because of too little transport capacity available (see above).

Today feeder roads reach even the remotest parts of the subcontinent. Many villages are still not connected by road, sometimes even not by a footpath (people have to balance on the little mud walls that separate the fields), but metalled roads will not be very far away. Only in times of extreme scarcities and massive food imports harbours may become transport bottlenecks. But this has become less of a problem after all Indian coastal states have built and/or expanded their harbours.

Instruments of food security III:
Increasing production

The really impressive expansion of food production has been dealt with already. This is true for almost all major food items. As far as cereals is concerned, it can be seen from Table 3.3.2.2. Worth mentioning is the expansion of non-vegetarian food, even in India. Predominantly Hindu India consumes much more animal products per head of their population than predominantly Muslim Bangladesh. Most remarkable is the increase in egg and poultry meat production. Poultry farms producing on industrial lines have become a common sight all over South Asia. Dairy production also multiplied. In the case of milk this has to do with packing: Milk quality constituted a major problem until polyethylene bags were introduced. Packed milk sells at a premium over "fresh" milk, because the buyer can be sure, that the milk has not been adulterated, which has been a common and hazardous (because untreated water was added) practice.

At various instances South Asian countries have become exporters of major food items. This especially applies to Pakistan where surplus (and often high quality) rice is exported and (cheaper) wheat is imported. India has become the second largest rice exporter in the world. There is, however, no major food trade between the states of the region, at least not officially: The boundary between India

Table 3.3.2.2 Food production in South Asia 1961-2002, in thousand metric tons

Country/ years	Rice (paddy)	Wheat	Barley	Maize	Millet	Sorghum	Total
Bangladesh							
1961-63	14,555	39	19	5	39	1	14,658
1971-73	15,964	106	21	2	50	1	16,145
1981-83	21,177	1,052	10	1	66	1	22,307
1991-93	27,312	1,082	10	3	65	1	28,471
2000-02	37,345	1,706	4	10	56	1	39,121
Bhutan							
1961-63	38	5	2	50	3		101
1971-73	48	7	3	65	4		132
1981-83	59	10	4	82	7		167
1991-93	43	6	4	43	7		110
2000-02	48	15	4	63	6		142
India							
1961-63	52,939	11,282	2,796	4,493	7,834	8,992	88,336
1971-73	63,282	24,992	2,580	5,764	9,386	7,928	113,834
1981-83	80,234	38,853	2,051	7,123	10,317	11,578	150,156
1991-93	113,814	56,011	1,615	9,219	9,633	10,773	201,065
2000-02	129,310	72,621	1,431	11,972	9,220	7,461	232,015
Nepal							
1961-63	2,108	137	20	845	65		3,175
1971-73	2,257	243	25	798	136		3,459
1981-83	2,383	553	23	744	119		3,822
1991-93	3,101	788	27	1,235	247		5,398
2000-02	4,170	1,200	31	1,470	278		7,149
Pakistan							
1961-63	1,707	4,003	120	499	385	246	6,960
1971-73	3,523	6,936	101	726	339	331	11,956
1981-83	5,107	11,731	173	983	250	223	18,466
1991-93	5,177	15,469	147	1,200	160	225	22,377
2000-02	6,457	19,443	105	1,666	198	223	28,092
Sri Lanka							
1961-63	999			10	20	1	1,030
1971-73	1,340			16	15	1	1,373
1981-83	2,290			24	15		2,329
1991-93	2,433			32	7		2,471
2000-02	2,783			30	4		2,817
South Asia (SAARC)							
1961-63	74,454	15,603	2,977	6,747	8,411	9,240	117,435
1971-73	88,671	32,284	2,730	7,371	9,930	8,261	150,358

Table 3.3.2.2 Food production in South Asia 1961-2002, in thousand metric tons—cont'd

Country/ years	Rice (paddy)	Wheat	Barley	Maize	Millet	Sorghum	Total
1981-83	113,633	52,199	2,261	8,957	10,774	11,802	201,069
1991-93	154,981	74,144	1,830	12,967	10,366	10,999	265,290
2000-02	184,283	96,185	1,606	16,681	10,040	7,685	316,485
Afghanistan							
1961-63	319	2,168	378	704	20		3,560
1971-73	390	2,355	355	717	30		3,846
1981-83	368	2,389	281	668	31		3,736
1991-93	312	1,772	222	397	22		2,725
2000-02	242	1,918	248	191	21		2,620
Myanmar							
1961-63	7,427	18		65	47		7,564
1971-73	8,045	31		58	39		8,183
1981-83	14,269	124		252	151		14,811
1991-93	14,936	135		201	134		15,419
2000-02	21,708	100		516	170		22,509

Notes: Three years averages, – Cereals include wheat, rice (paddy), barley, maize, rye, oat, millet, sorghum, buck weat and others. Cereal production on the Maldives is negligible, data for Bhutan and Afghanistan not always consistent.
Source: FAOSTAT Database Results (http://apps.fao.org...), May 21 2003.

and Bangladesh runs over open plains through mostly densely populated areas; whenever prices differ in India and Bangladesh large quantities of food items are transported across the border unchecked and unaccounted for.

Instruments of food security IV: Imports and food aid

From the 1940s until the 1970s the United States and Canada (plus Australia and later the European Community) were the major sources of imports whenever there was a need. As a reaction to the Great Depression and the Second World War most countries had introduced a system of price stabilisation and support for their major agricultural products. The response was so good, that large stocks could (or had to be) built up at a time when population growth accelerated in the then emerging Third World. For India and Pakistan, in particular, doomsday scenarios of Malthusian proportions were feared to become reality. The United States already had helped to reconstruct Europe. It only seemed natural, and humanitarian, now to use agricultural surpluses as wheat loans (to be paid back in kind – they were turned into outright grants later). It also seemed to be an appropriate measure to stem the "red tide" of communism in Asia. It was the time, when China "was lost" and other countries were bound to fall according to the domino theory. Throughout the 1950s and 1960s India and Pakistan imported millions of tons of food-grains on an almost yearly basis. The dependency was felt only in 1965, a year of an exceptionally poor harvest, when India and Pakistan could be forced by the USA (together with the Soviet Union) to end their (short) border war. This would have happened again in 1971, when India and Pakistan were again at war, if food production had not risen dramatically as a result of the green revolution of the late 1960s.

Bangladesh, however, had to feel the force of "food power", when they became subject to the Cuba embargo of the USA. According to this embargo, countries were excluded from US aid, if they did trade with Cuba. Bangladesh during the first years after Independence followed an economic policy model close to that of India, the country that had helped then East Pakistan (Bangladesh) to free themselves from (West) Pakistan domination. India, for this reason, was heavily leaning on the Soviet Union. It was the time of the final stage of the Vietnam War and the height of Socialist policy in India. The economy of Bangladesh at that time was going from bad to worse. Raw jute and jute products were the only export goods, the market for both of them stagnated and any export order was welcome. The export of jute sacks to Cuba presented the USA with a golden opportunity to discipline the Bangladesh government; a (possible) waiver was not granted. The Bangladesh government on their part had underestimated the extent of the bad harvest of that year; distress calls were sounded very late and the result was a famine, which cost about 50,000 lives (Faaland 1981). It was the last major famine in South Asia (besides the five year drought in Afghanistan, which coincided with the Taliban regime). When the USA finally resumed their aid to Bangladesh, relief arrived at the time of the next (better) harvest, and had the typical price depressing effect.

As a result of the bad experiences with food imports, i.e. the Great Bengal Famine 1943, the Indo-Pakistan war of 1965 and the Bangladesh Famine of 1974, food imports/aid are discussed in South Asia as a political, not an economic problem. On the other hand, the green revolution saved India from any food related pressure in 1971. Ever since large food-stocks made India immune to "food power": it could manage the drought of 1987, had more negotiation power in the foreign exchange crisis of 1991, and after the economic sanctions in response to the nuclear tests of 1998. The drought of 2002 resulted in the most serious fall in production since decades, without threatening the (macro) food basis.

But there are also economic reasons, not to rely too much on the world market. The main argument is that India is the second largest consumer of food-grains (after China) in the world. Its food-grains consumption has almost the same size as the whole world market. If a shortfall of 30 m tons, like the one of 2002, had to be met by the world market, significant price rises are to be expected. This all the more, since markets today react instantly. Satellite images have improved market intelligence, traders are well aware of any presumptive increase in demand.

The unprecedented high foreign exchange reserves that India presently enjoys (and many other countries, thanks to record current account deficits of the USA) could erode very fast, as the example of other countries (e.g. during the "Asian crisis" of 1997) shows, they are, thus not a guarantee that foreign exchange will be available if needed for food imports. Srinivasan and Jha (1999) have shown with the help of a stochastic dynamic simulation model that variable levies on trade turn out to be superior compared to buffer stocks in stabilising prices under liberalised trade, although more to the benefit of consumers vis-a-vis the producers. This even holds true despite the fact that domestic prices are less volatile than international prices.

Instruments of food security V: Price policy and market intervention

Governments, especially in India and Pakistan, have a long tradition of market and price intervention. This applies not only to state procurement and state trading for major food items (see above), but even more for intervention in the factor markets. Explicit and even more implicit subsidies have become powerful tools to stimulate production. Under conditions of a mostly arid or semi-arid climate, water is as important a production factor as land. Besides on-farm irrigation like the traditional Persian wheel, canals and tanks have been major sources of irrigation, often limited to the post-monsoon season. From the mid 1800s onwards the British built a series of barrages over the major rivers and distributed the water through an extensive network of canals to the fields. Such irrigation opened new areas for agriculture, like the "canal colonies" in the Punjab. Settlers were drawn from other areas, land also was distributed to army and government personnel. Agricultural production was increased and population pressure relieved *uno actu*. But irrigation was not free of charge. Water taxes or "cess" (in the north: *abiana*) were substantial and could cost up to a third of the production value. During the great depression such taxes were reduced as well as the land tax, and after independence the landlords, who dominated much of politics (especially in West Pakistan), saw that agricultural taxes were virtually abandoned, until the water tax yielded hardly enough revenue to cover collection cost. After Independence large dams were built and water stored in huge reservoirs, allowing a much more secure water supply, which is so essential for high yielding varieties. These cost had to be borne by the exchequer and – to some extent – by foreign aid. Tube-wells, the major source of ground water mobilisation, use pumps, which are driven by electrical or diesel motors. Such on-farm investment was financed by the government owned/controlled financial institutions at preferential conditions. The recovery rate of such loans has been poor. Especially the larger debtors/land owners did not bother to repay their loans or even any interest, expecting – for good reason – that such loans might be written off (loan holidays) and/or new loans granted. In addition to this diesel and especially electricity was – and still is – provided at highly subsidised prices and in some parts of India is even free of charge.

Such policies are, of course, financially unsustainable. The budget deficit in India already has reached the ten percent (of GDP) mark. The chief minister of the Indian state of Andhra Pradesh successfully campaigned for raising water and electricity rates and even won his elections. This is a clear indication that voters know that free riding would be harmful on the longer run. Other governments have not followed as yet. Heavy subsidies on agricultural inputs, tax exemptions, tax evasion and non payment of liabilities deprive the state (and the parastatal or semi-government agencies) of the funds urgently needed to built up the infrastructure and to provide credit for on-farm investment.

Instruments of food security VI: Raising incomes and purchasing power

Table 3.3.2.1 shows that per capita availability and consumption has increased in all South Asian states over the last four decades. The only exemption may be Bangladesh, where the per capita calorie intake has reached the level of the late 1960s, i.e. the years before the civil war. If we look at the composition of food, even here some improvement can be detected, at least as the intake of fats is concerned, which rose by almost one half. Levels are, however still very low by any standard. There are no up-to-date FAO figures available for Bhutan and according to the older ones, Bhutan has about the same low level of consumption as Bangladesh. But older figures for Bhutan suffer from highly inflated populated figures, the actual consumption per capita may be much higher.

This certainly is not the case for Afghanistan. The last figures available, i.e. for the mid 1990s, show a very low level of consumption, much lower than in any country of the region and much lower than before the pre-war times. But any figures for the last two decades are nothing more than intelligent guessing. Standard sources either repeat older, outdated figures or work with trend extrapolations. Under the condition that the political situation remains stable and/or can be improved, there may be much better data available within the next one or two years. A plan for an agricultural census was announced in December 2002.

India, Pakistan and Bangladesh were able to improve their food supply. The same holds true for Nepal, and to a lesser extent for Sri Lanka, which suffered from the long civil war in parts of the country. The greatest leap forward, however, was experienced by the Maldives, which once were the worst fed in the region and now are the best fed: The country has become a favourite tourist destination and enjoys the highest (average) per capita income of all South Asian states. Almost all food is imported.

The costs of food security

Economists emphasise, that costs are to be measured in opportunities foregone. If food security is measured in lives (or life years) saved, then we might expect, that no price would be too high. But there are convincing arguments, that not all public money spent under the headings of food security and food production does actually help to save lives. On the contrary: If it is true that much of the money does not secure anybody's food supply, there might be substantial opportunity costs in the form of money not spent on safe health, drinking water and other public amenities, not to speak of education.

In order to stimulate agricultural production, various subsidies have been paid, especially in India, Pakistan and Bangladesh. As for India, the major explicit one is on fertiliser. This one, however, aids a partly inefficient fertiliser industry rather than stimulate fertiliser consumption. As a result, the fertiliser consumption per area unit is stagnating and is even lower than in Pakistan and Bangladesh. Among the major implicit (hidden) subsidies, those on (canal) irrigation water, on electricity (for powering electric tubewells) and agricultural credit (e.g. for tractors) are the most important. (Canal) Water is not metered and provided free of cost or charged at flat rates. Accordingly water is overused, the cropping patterns do not reflect (economic) scarcities: crops that need a lot of water are grown at locations of high potential evapotranspiration and seepage, whereas high protein dry crops like sorghum, millet and pulses are more and more marginalised. Similarly electric power for agriculture is not metered and provided at flat rates or even free. Official figures show that one quarter of electricity consumption in India is for agriculture/irrigation. These figures, however, are questioned, because not measuring consumption is an open invitation for fraudulent practices, which at best are shown under "system losses". As in the case of water, power is subject to over-utilisation (Chopra 2003). The effects are power shortages, extreme voltage fluctuations, load shedding and power cuts, resulting in high fixed and running costs for standby-arrangements (generators, diesel pumps) and wear and tear of the equipment. Revenue income from water and power is low and covers, at times, just the costs of collection. All other costs have to be met by the exchequer. No wonder, that the Indian budget deficit has reached alarming proportions (see above). As for agricultural credit India has provisions for priority sectors, basically a system of cross subsidising agricultural credit at highly preferential conditions at the expense of the other sectors. Many of these credits are "not performing", i.e. neither amortisation nor interest is paid by the creditors without any consequence. On top of it, there have been instances of loan waivers (loan *melas*); the defaulters even managed to get new loans. Most of the banks are in the public sector; the total amount of "non performing assets" must be substantial and not easily to assess (e.g. if loans are "repaid", but actually rolled over).

Government procurement has been so "successful", that India is now holding the largest food-grains reserves in the world. Even after the poor harvest of 2002 the buffer stocks are at least twice as high as considered to be needed for emergency. The quality of the stocks, however, has been widely questioned in India (Raghavan 2003, Report of the High Level Committee 2002). Given the fact, that around a fourth of the Indian population lives below the (national) poverty line, which is defined as income sufficient to meet minimum food requirements, the buffer stocks today serve more the purpose of stabilising/raising producer prices and incomes than securing food supply.

The public food distribution system (PFDS) suffers from its urban bias, as the Indian government consents in their economic survey (Tyagi 1990). Some authors go so far as to state, that the poor are rather hurt than helped by the system, for example when the government procurement drives up prices in the country side and poor farmers with no access to the PDS have to pay a higher price than without procurement (Ramaswami and Balakrishnan 2002, Shankar 2002, Swaminathan 1995).

Outlook

Population growth will continue at least until the middle of the century. By then India may have surpassed China as the most populous nation, Pakistan is expected to rank third among all countries of the world, then. If a healthy economic growth, especially in India, can be maintained, food demand will increase much faster than population growth. It should be possible to meet such a higher demand by own production, since yields per area unit are still low in South Asia by international standards. Increasing production, however, requires a steady supply of inputs like water, power, fertiliser, pesticides and know how. Alternatively more food could be imported if India and the other South Asian states would decide to globalise their economies, i.e. to integrate into the global system of division of labour.

References

Blyn G (1966) Agricultural trends in India, 1891-1947. University of Pennsylvania Press, Philadelphia

Bohle HG et al. (eds) (1997) Ernährungssicherung in Südasien. Siebte Heidelberger Südasiengespräche. Beiträge zur Südasienforschung 178, Franz Steiner, Stuttgart

Chand R (2002) Government intervention in foodgrain markets in the new context. National Centre for Agricultural Economics and Policy Research, New Delhi

Chand R (2003) Policy and technical options to deal with India's food surpluses and shortages. Current Science 84(3):388-398

Chopra K (2003) Sustainable use of water in India: The next two decades. Fourteenth Dr. Kanwar Sain Memorial Lecture. Institute of Economic Growth, New Delhi

Chopra RN (1981) Evolution of food policy in India. Macmillan, New Delhi

Chopra RN (1988) Food policy in India. A survey. Intellectual Publishing House, New Delhi

Chopra RN (1986) Green revolution in India. The relevance of administrative support for its success. Intellectual Publishing House, New Delhi

De Janvry A, Subbarao K (1986) Agricultural price policy and income distribution in India. Studies in economic development and planning 43. Oxford University Press, Delhi

De Vylder S (1982) Agriculture in chains. Bangladesh: a case study in contradictions and constraints. Zed, London/Vikas, New Delhi

Dréze J (ed) (1999) The economics of famine. Edward Elgar, London

Dréze J, Amartya S., Athar H. (eds) (1995) The political economy of hunger. Clarendon, London

Faaland J (ed) (1981) Aid and influence. The case of Bangladesh. Macmillan, London/Basingstoke

FAO, FAOSTAT. http://www.fao.org

F. E. Division, Directorate of Economics and Statistics (ed) (2002) Bulletin on Food Statistics (1998-2000). F. E. Division, Directorate of Economics and Statistics, Department of Agriculture & Cooperation, Ministry of Agriculture, Government of India, New Delhi

Jalan B (1997) India's economic policy: preparing for the twenty-first century. Penguin, New Delhi

Kapur, D, Patel UR (1993) Large foreign currency reserves: insurance for domestic weaknesses and external uncertainties? Economic and Political Weekly, 28(11):1047-1053

Knight H (1954) Food administration in India, 1939-1947. Stanford University Press, Stanford

Lele UJ (1971) Food grain marketing in India: private performance and public policy. Cornell University Press, Ithaca

Malthus TR (1982, 1798,1830) An essay on the principles of population and a summary view of the principle of population. Edited with an introduction by Anthony Flew. Penguin English Library

Raghavan M (2003) Food stocks: managing excess. Economic and Political Weekly 38(9):873-875

Ramaswami B, Balakrishnan P (2002) Food prices and the efficiency of public intervention: the case of the public distribution system in India. Food policy 27: 419-436

Ravallion M. (1987) Markets and famines. Clarendon Press, Oxford

(2002) Report of the high level committee on long-term grains policy. Department of Food and Public Distribution, Ministry of Consumer Affairs, Food and Public Distribution, New Delhi

Sen A (1994) Poverty and famines: an essay on entitlement and deprivation. Oxford University Press, New Delhi

Shankar K (2002) Starvation and deaths in UP and PDS. Economic and Political Weekly 37(42):4272-4273.

Singh KS (1992) People of India: an introduction. Anthropological Survey of India. Oxford University Press

Smith LC, Haddad L (2000) Overcoming child malnutrition in developing countries: achievements and future choices. Food, agriculture, and the environment discussion paper 30. International Food Policy Research Institute, Washington, D.C.

Srinivasan P V, Jha S (1999) Food security through price stabilisation. Economic and Political Weekly 34(48):3299-3304

Swaminathan M (1995) Revamped Public Distribution System: a field report from Maharashtra. Economic Political Weekly 30(36):2230

Tyagi DS, Vyas VS (eds) (1990) Increasing access to food: the Asian experience. Sage, New Delhi

Tyagi D S (1990) Managing India's food economy: problems and alternatives. Sage, New Delhi

Zingel WP (1999) Genug Nahrung für eine Milliarde Inder? In: Draguhn W (ed) Indien 1999. Institut für Asienkunde, Hamburg, pp 217-233

Zingel WP (2003) Nahrungssicherungspolitik in Indien. In: Draguhn W (ed) Indien 2003. Institut für Asienkunde, Hamburg

Part 4

Scientific Challenges for Anthropogenic
Research in the 21st Century

4.1 Scientific Challenges for Anthropogenic Research in the 21st Century: Problems of Scale

Rik Leemans

Environmental Systems Analysis Group, Wageningen University, P.O. Box 9101, 6700 HB Wageningen

Introduction

Research into the behaviour of the earth system in the recent past, the current era and the near future, a period that nowadays is recognised as the Anthropocene (Crutzen 2002), requires an adequate description of the dynamics of many different components and their interactions. Major components are the atmosphere, the biosphere (including humans and their societies) and the oceans. The dynamics of all these components are nowadays influenced by human activities. The natural components and the anthropogenic components of the earth system should be studied in an integrated way. In this chapter I will provide ways to actually link the natural or environmental and anthropogenic dimensions.

Most components function on many different scales and levels: from molecules or other particles to the whole atmosphere, from individuals to nations, from chlorophyll grains to whole plants, from plants to ecosystems, from landscapes to biomes, from streams to watersheds to the full hydrological cycle, etc. Many processes determine the specific behaviour of each component. These processes determine linkages and interactions at different scales and levels between different components. Such complexity should be represented in global change research because one can rarely describe complex behaviour at just one specific level. Interactions and feedbacks across scales thus strongly determine earth system dynamics (Schellnhuber 1998).

Unfortunately, 'scale' is one of the most abused notions in earth system research. Originally, when solely used for cartographic mapping purposes, scale was well defined and indicated smallest detail that still could be displayed. Large scale maps display many details, while small-scale maps display few. Currently, scale is more and more used to indicate the duration, extent and level of different spatial, temporal or organisational features. Fine scales display spatial, temporal and organisational detail, while coarse scales cover large regions, periods or whole nations.

In order to avoid confusion, I'll use this 'fine' and 'coarse' terminology throughout the chapter and leave "small scales" and "large scales" solely for mapping purposes.

Scale refers to the dimensions, in space, time or organisational level, of a particular object, system or phenomenon (O'Neill and King 1998). These are measured by the appropriate sub-components (i.e. the unit of analysis) and expressed in obvious units, such as meters or years. Scale has three major components: extent or duration, resolution, and grain (Blöschl and Sivapalan 1995; Figure 4.1.1). The extent or duration defines the boundaries, the area or the magnitudes. Resolution defines the finest detail that is distinguishable. The grain is the finest detail that is internally homogeneous. In principle, scales related to the physical and temporal dimensions are continuous and resolutions are generally selected in a very pragmatic way depending on, for example, data availability. The physical dimensions define data structure, which is generally spatially or temporally explicit. Organizational scales are generally discrete (e.g. individual, family, community, municipal, province, country) and not continuous. Even the grain could vary (e.g. small and larger countries). Organisational data is available in a tabular format for the specific level. The word 'level' is used here to describe the discrete scales of social organisation.

Scale is also closely related to predictability: fine-scale events show more variability than coarse scale events do. This is because the effects of local heterogeneity are averaged out at coarser scales, so that patterns become more predictable (Levin 1992). On the other hand, studies focussing on broad-scale patterns lose

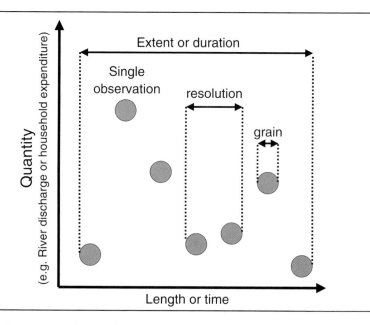

Fig. 4.1.1 Components of scale of observations

predictive accuracy at specific points in space and time (e.g. Costanza and Maxwell 1994). Determining the optimal scale and interactions with lower and higher scales or levels of your specific global change research field is one of the major challenges and depends on the scientific research question(s) or the objective(s) of the study.

This chapter deals with scale issues. First, I will further discuss why scale matters. Then, I will present some empirical, modelling or assessment studies that integrated processes along different dimensions and different scales. Finally, I present and discuss the different dimensions and related scale levels required in comprehensive analysis of the earth system.

The dynamic behaviour of the earth system is central in this discussion. Many other studies have emphasised the behaviour of individual components and the factors that influence them. These factors are termed drivers. I will especially focus on the interactions between drivers and the different components of the earth system. This illustrates the complexity of earth system analysis and explicitly highlights the importance of scales at which processes and drivers operate and interact. It further allows improving the integration of natural (e.g. ecological, chemical and physical) and human (e.g. economic, social and cultural) processes. This provides adequate guidelines of how to deal with scale issues in global-change studies. Unfortunately a comprehensive earth system science with proper illustrative examples has not yet emerged. Therefore, I will present many examples from my own discipline land-use and ecosystems research.

Why scale matters?

In studying earth system dynamics, scale matters for two main reasons. First, physical, geochemical, ecological and social systems and processes operate at a wide variety of spatial and temporal scales from very fine to coarse and from short to extremely long. While crossing scales, they can change their nature and sensitivity to various driving forces. Results obtained at one scale are not automatically valid at another (Kremen et al. 2000; McConnell 2002). Thus if processes are observed or assessed at scales significantly finer or coarser than their characteristic scale, misleading or erratic conclusions can easily be drawn. For example, it is inappropriate to draw any conclusions regarding long-term trends based on short-duration time series data. People do not infer that the primary productivity of the world is declining just because in the Northern Hemisphere leaves die in the autumn; based on experience, it is obvious that this is part of a longer-term seasonal cycle. Nor can it be assumed that because a change is occurring at one location it is occurring equally at all locations. It is springtime in the Southern Hemisphere during fall in the north.

Second, cross-scale interactions exert a crucial influence on outcomes at a given scale. Focusing solely on a single scale can miss these interactions. Looking at a particular issue top-down, from the perspective of coarser scales or higher organisational levels, can lead to different conclusions than looking at the same issue bottom-up, from the perspective of finer scales or lower levels (Berkes 2002, Lovell et al. 2002). The scale of the analysis influences both the framing of an issue and the range of possible actions and institutional responses. Where cross-scale interactions between components of the Earth system occur, there should be no expectation of finding a single most appropriate level for response or policy. In most cases, mutually supportive policy changes and responses at different levels are required in order to bring about desired results.

Additionally, some specific aspects of spatial and temporal scales are important in characterising earth system behaviour. The first is related to temporal interactions. Not all processes evolve at similar rates, which could lead to time lags in responses. Events that happen can easily be triggered by (combinations of) factors that occurred in the past. Additionally, grains at a specific spatial scale are generally interrelated. Such spatial autocorrelation could well confuse the analysis.

A critical appraisal of the unit of analysis, its characteristic scales and its temporal in spatial interactions is thus urgently needed in each study of (components of) the dynamic behaviour of the Earth system.

Interactions and scales

Within the context of global change studies one is often confronted with systems composed of many intricate interrelated elements. Such systems are complex systems. Complexity, however, is by no means synonymous with non-linearity: it also involves connectivity. The complexity arises from interactions between components, such as synergies and feedbacks, that operate on different scales and levels. Synergistic effects in the behaviour of the earth system may not only arise from the interplay of endogenous factors but also from other driving forces directly related to human activities.

For example, the complex interactions of processes related to the exploitation of forests, droughts and fire are observed to constitute a serious threat to tropical forests. Logging increases forest flammability by reducing leaf canopy coverage and allowing sunlight to dry out the forest floor. This permits fires ignited on agricultural lands to penetrate logged forests. Once a fire has entered the forest, a feedback loop is initiated which involves increased susceptibility to future burning, fuel loading and fire intensity (Nepstad et al. 1999). These insights are of great concern since various climate models predict a substantial decrease in precipitation in tropical forests areas (Cox et al. 2000; Leemans 1999) and increase in frequency and intensity of ENSO events. Forest impoverishment through logging and fire causes a significant release of carbon to the atmosphere. It has been calculated that if just one-fifth of the Roraima region in Gyana, which was vulnerable to fire in the 1998 dry season, had caught fire, net carbon emissions would have risen to 10% worldwide emissions (Nepstad et al. 1999).

Other clear examples of the importance of interactions in ecological behaviour are given by Scheffer et al. (2001). They show that gradual changes in drivers, such as nutrient loading or habitat loss could initiate dramatic sudden changes in ecosystem response, which seem often to be irreversible. And when reversible, a switch towards the original status was far below the level at which the other switch occurred. The pathways between the different states strongly differed. Scheffer et al. (2002) applied this understanding to further develop economic resource-use models to determine sustainable use on the short and longer term.

As has been shown in several recent papers (e.g. Claussen 1996; Ganopolski et al. 1998), prominent features of Holocene climate and vegetation displayed by paleodata can only be understood by taking into account the complex suite of interactions between atmosphere, ocean and vegetation. Paleodata suggest that in the Mid-Holocene (9 to 6 thousand years B.P) Earth's climate was quite different from that of today. Summer in the Northern Hemisphere was warmer and boreal forests extended north of modern tree line. In North Africa, climate was wetter and vegetation covered a substantial part of the Sahara. Transition from Mid-Holocene to modern climate was triggered by a smooth change in the Earth's orbit and the tilt of the Earth's axis. Accounting for a strong positive feedback between vegetation and precipitation can only reproduce the transition in the Sahara from a "green" to a desert equilibrium, which was found to have happened rather abruptly. And the cooling of the Northern Hemisphere, with consequent southern retreat of boreal forests, is shown to have been amplified by the sea-ice albedo feedback superimposed on the vegetation-snow-albedo feedback.

Complex feedback and synergetic interactions cross scales are thus important for determining the behaviour of (components of) the earth system and for projecting future changes in that behaviour under influence of human activities.

Traditional treatment of scale in earth system analysis

In determining the relevant scale of any earth system study one must first define the relevant dimensions, their relevant grain and resolution and then the level or levels at which interactions occur. This actually requires an advanced understanding of the systems studied because processes that operate at particular scales are related to processes at other scales as well. Trying to identify and quantify cause and effect has been an important endeavour in global change research.

During the late 1960s and early 1970s, debates began about the factors that lead humans to have adverse effects on the biophysical environment. A number of so-called "root" causes were asserted: religion (e.g. White 1967), common property institutions (e.g. McCay and Jentoft 1998), capitalism and colonialism (e.g. O'Connor 1988). None of these hypotheses of single dominant causes, however, could sustain empirical scrutiny and thus explain environmental change.

The empirical IPAT formulation (Impacts = Population * Affluence * Technology) was an initial attempt by Ehrlich and Holdren (1971) to move beyond simple arguments about single causes by acknowledging that:

- there are multiple human drivers of environmental change;
- that their effects are multiplicative rather than additive;
- that increases in one driver can sometimes be mitigated by changes in another driver; and
- that assessing the effects of human drivers requires both theory and empirical evidence.

Dietz and Rosa (1994) provide a well-documented history of IPAT and related arguments about such interrelated factors.

IPAT itself continues to be used in discussions of the drivers of environmental change (Waggoner and Ausubel 2002) and the IPAT accounting framework finds productive use in industrial ecology (Chertow 2001). However, formulations that build on but move beyond IPAT are emerging rapidly in global and regional models. The importance of population and affluence on consumption continues to be examined. A variety of studies demonstrate that population size has an effect on impact but sometimes is less important than other factors, especially locally and regionally (e.g. Palloni 1994; York et al. 2003). A substantial literature examines the effects of affluence on environmental impact (reviewed by Stern 1998), including a number of analyses that suggest that such effects are highly context dependent (Roberts and Grimes 1997).

Over the last decade adding additional factors, such as specific socio-political, biophysical and cultural drivers, has further refined the analysis. But these top-down approaches to understand, explain or project the environmental change still rely heavily on those highly aggregated drivers, whose value has recently been questioned (e.g. Lambin et al. 2001; Myers and Kent 2001; Young 2002). The most important recent advance in our understanding is the elucidation of a broader variety of interacting drivers that become more important in the local context.

The individual importance of global drivers thus cannot be assessed in a simple way. There is no clear hierarchy of drivers that encompass cause and effect because of large regional differences. Individuals and societies try to influence their environment and fulfil their needs by evaluating expected outcomes. If undesired impacts are foreseen, mitigating decisions can be made. This approach is made operational most clearly in the Driver-Pressure-State-Impact-Response (DPSIR) scheme that was originally developed by the Organisation of Economic Co-operation and Development (OECD's InterFutures Study Team 1979). Drivers are any natural or human-induced factor that indirectly causes a change in any system. The pressures directly cause changes in the state of a system. Impacts are the consequences of these state changes. The response includes the activities to ameliorate impacts by reducing pressures through modifying the drivers. The DPSIR scheme linearises complex chains of cause and effect. Unfortunately it is often difficult to unambiguously determine whether a factor is a driver, a pressure, a state or an impact variable. This strongly depends on one's research or assessment perspective.

Many assessments, however, have followed the DPSIR approach. For example, the Intergovernmental Panel on Climate Change (IPCC 2001) structured its assessment along these lines (activities; emissions; concentrations; climate change impacts; mitigation and adaptation responses; c.f. IPCC 2001), recognising that responses in turn alter activities (through mitigation measures) and impacts (through adaptation measures). The natural science assessment of IPCC strongly focused on the state aspects (How do we estimate the level of climate change and its impacts from emissions). Unfortunately, the use of such DPSIR approach has neglected important feedbacks between drivers and major components of the earth system.

The Millennium Ecosystem Assessment (MA 2003) has taken a more comprehensive approach trying to explicitly address trade-offs and synergies. The MA conceptual framework is therefore a closed loop and displays different interactions between drivers, ecosystems, ecosystems services and human well being (Figure 4.1.2). Two types of drivers are recognised: direct and indirect. A direct driver unequivocally influences ecosystem processes and can therefore be identified and measured to differing degrees of accuracy. An indirect driver operates more diffusely, often by altering one or more direct drivers, and its influence is established by understanding its effect on direct drivers. Additionally, the MA recognises that there are cross scale linkages. Determining tradeoffs and synergies between different decisions and other responses will be a central theme in the MA. This requires that the assessment takes a close look at the interactions of drivers at specific scale levels and

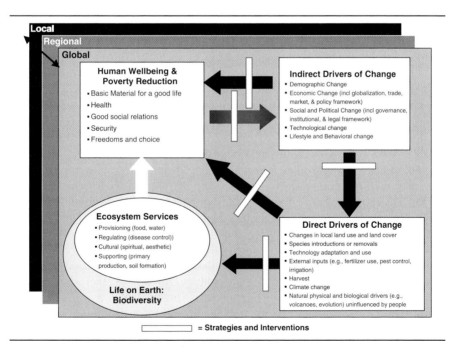

Fig. 4.1.2 The conceptual framework of the Millennium Ecosystem Assessment

between spatial, temporal and organisational dimensions. A strong multidisciplinary team from both the social and natural sciences therefore executes this assessment, which will become available early 2005.

The earlier earth system models (Rotmans 1990) denied such complexity and globally aggregated the dynamics of both human and environmental processes. Although such aggregated models provided some insights (e.g. global greenhouse gas emissions are nowadays dominated by fossil fuel use and not by land use), the required policy measures could never be adequately simulated because policies do not operate globally. This is one of the reasons why these highly aggregated models are being retired nowadays. Many recent earth system models have much more regional detail (e.g. Alcamo et al. 1998; Kainuma et al. 2002).

Local and regional detail is nowadays always added for a better representation of the interactions at different scales and levels. For example, in the earlier model, deforestation was modelled as a change in global extent. The globally average carbon content (a highly calibrated estimate) in forests determined the resulting CO_2 flux from deforestation. The actual complex set of causes of deforestation (see Geist and Lambin 2002) was neglected. The current models account for different causes and heterogeneity in carbon storage in forested ecosystems in order to estimate deforestation fluxes. The more advanced models, such as cellular automata (e.g. White et al. 1997; Costanza et al. 2002) even consider spatial and temporal interactions.

Improved treatment of scale in earth system analysis

Adding local and regional details is actually not enough to capture the behaviour of the different natural and anthropogenic components of the Earth system. Each component has characteristic scales and levels (Figure 4.1.3). The organisational levels of the anthropogenic dimensions span individuals, families, communities, and nations. All these levels have specific recognisable actors and institutions, each with its own choices, decisions and boundary conditions.

For example, an individual farmer from the Great Plains produces for the international markets. His incentives are strongly determined by international grain prices. His earnings are (partly) invested to provide a college education for his children. He uses technologies (e.g. tractors, high-yielding crop varieties and pest control), which is developed by large multinational companies and governmental research. The farmer's family uses fossil fuels to warm and cool their house, travel by car and planes and many other activities. This farmer also chairs a regional farmer union to lobby politically for lower taxes, crop-failure relief when droughts arise, and guaranteed access to foreign markets. Another farmer living in an African country produces crops to feed his own family and to provide some surpluses to sell at local markets to purchase simple tools and, if possible, fertiliser. His children help to work the land and shepherd the few cattle. The technology he uses is simple, evolved regionally and adapted to local conditions. The energy he uses stems

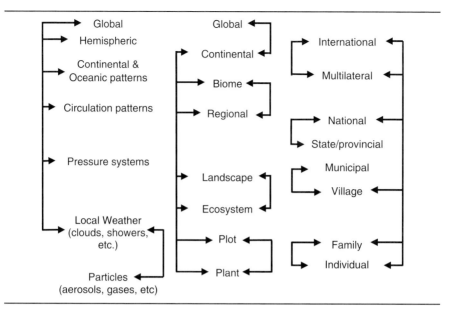

Fig. 4.1.3 Different scales and levels in the atmosphere (left), ecosystems and land (middle) and society (right). The diagram is illustrative in indicating that different scales can be distinguished in different domains.).

from locally collected fuel wood. This farmer is not organised and strongly dependent on decisions made at communal and national levels.

Both examples show that there are strong cross-scale linkages that determine the behaviour of individuals. Some scales can be influenced directly by them. They can decide how to use their private resources. They can plan and time their activities (e.g. required to grow crops). Features at higher scales (e.g. price of crops and fertiliser) cannot be influenced by these individuals but seem emergent properties of these higher levels. At national levels decisions are made to accelerate the adoption of certain activities and/or slow down others. These levels thus determine many of the boundary conditions for the lower levels. Sometimes, well-organised individuals can influences such policies at these levels, but rarely as an individual.

The highest organisational level has it own dynamics. Internationally, there are many types of collaborations between countries. There are bilateral agreements on issues like trade, cross-boundary air pollution or scientific exchange. There are strong collaborations from groups of countries (e.g. NAFTA, EU, ASEAN). The advantages of these multilateral collaborations are generally economies of scale with improved exchange of people, goods and capital. Finally there are the international United Nations (UN) conventions, such as the Framework Convention on Climate

Change (FCCC) and the Convention on Biological Diversity (CBD). The decisive bodies of these conventions are international institutions but operate not as single entities, like a nation, a family or individual. The conventions are made up of countries, which in the conference of parties take decisions often on basis of consensus. This means that the country or national level still dominates at this international level (Lambin et al. 2002). Additionally, the way the international conventions are implemented is that they do not develop specific policies. They define targets and negotiate instruments, such as the clean development mechanism or joint implementation in the FCCC, to allow individual countries to effectively develop policies to reach targets.

The environment is altered by human activities in many different ways. However, the linkages between these activities and the environment are not always clear. Figure 4.1.3 also show the different scale levels in ecological systems (ranging from plants to biomes) and atmospheric or climate systems (ranging from molecules to coarse scale circulation patterns). The link between anthropogenic organisational levels and the environment can be complicated. For example, land use and land-use change patterns emerge from land-related activities of individuals. These activities are always local and involve management/cultivation of plants at the field or plot level. Communities or nations do not change land use directly, only indirectly by changing the drivers that alter the behaviour of individuals. They thus define the boundary conditions of the local decision makers. This means that the proximate drivers of land use change (i.e. the actual land use activities, such as creating fields, ranging, cropping) always operate at the lower or local scales. This local focus leads to the enormous heterogeneity in human land uses and landscapes all over the world.

Land use change within these local landscapes change ecosystems in various ways. Agricultural systems generally store less carbon, so when land is converted from the original vegetation carbon is emitted to the atmosphere through slow processes, such as enhanced decomposition, and fast processes, such as burning. It is the cumulative effects of land use changes over large areas that make a significant contribution to total land use related emissions, which alter the global composition of the atmosphere. Although the mechanisms are different, biodiversity (i.e. the biological diversity of genes, species and ecosystems) decline is also a cumulative consequence of land use change. The mechanisms here are habitat destruction and fragmentation, introduction of invasive species, pollution and other stresses stemming from multiple sources some of which can be moderated by appropriate conservation measures. Biodiversity decline is considered an important aspect of global change because the threats to biodiversity increase everywhere (Heywood and Watson 1995), while conservation measures seem less and less effective. Land use change thus contributes foremost through local changes of ecosystems to global change.

Climate change is a truly global feature. The atmosphere takes up emissions from land use (locally determined) and energy use (locally and regionally determined), where these emissions are rapidly mixed (within a year for a single hemisphere, within a few years even between both hemispheres). Only seasonally distinct latitudinal patterns can be distinguished but globally and annually atmospheric concentrations of greenhouse gases increase gradually (Prentice et al. 2001). These globally homogeneous concentrations alter atmospheric circulation patterns, which in turn change local and regional climates and weather patterns. Such climate change is again locally specific (especially precipitation) and influence locally ecosystems and human activities.

The linkages between anthropogenic and environmental processes occur not at single scales or levels. The interactions between different domains happen at various levels. For example, the causal chain 'climate change' operates from local, global and back to local again. The constraints of the land-use change, in contrast, are set by coarser scales (such as the communal and national levels or regional weather patterns) and act mostly at the local level. Its impacts become cumulatively important for other domains.

Concluding remarks

The spatial heterogeneity, diversity and variability in the environment results from all these interacting scales. Modern tools, such as integrated assessment models, geographic information systems and decision support systems nowadays can capture this quite well. Recent advances in integrated assessment modelling (e.g. Alcamo et al. 1998; Stafford Smith and Reynolds 2002) and comprehensive analyses of environmental problems (e.g. Petchel-Held et al. 1999; Ostrom et al. 2002) have also shown that analysing causes of environmental change requires a multi-scale and multi-dimensional assessment of major components of the system, their dynamics and interactions. This understanding is leading to a rapid advancement of different approaches that integrates cross scale environmental processes and the behaviour of different actors at different levels.

A better appreciation of the feedbacks, synergies and trade offs among these components in the past improves our understanding of current conditions and enhances our ability to project future outcomes and policy options. However, this is easier said then done because it involves a clear analysis of the scales at which major processes operate. Empirical research still strongly emphasises correlations at single scales leading to models that do not capture realistic behaviour and interactions, especially at smaller scales. Global change research must amalgamate major processes in the main system components by emphasising major linkages at their characteristic scale.

The history of global research moves away from the emission scenarios and impacts assessments of **climate change**, which focus on regional monitoring and prediction of climate change, impact modelling and socio-economic analysis of mitigation and adaptation strategies, towards more realistic **global change** research, which focuses on the interdependency and synergies of global and regional processes, to end in **sustainability research**, which integrates the socio-economic driving forces to explore possible development trajectories with a significantly smaller environmental impact. This apparent move requires an understanding of the complex mechanisms underlying climate change, global change and sustainability and the interactions between driving forces, impacts and responses that define the future of the earth. Spatial and temporal scales in the environment remain important, but comprehensively and functionally linking them to the various organisation scale levels will be the major challenge.

References

Alcamo J, Leemans R, Kreileman GJJ (1998) Global change scenarios of the 21st century. Results from the IMAGE 2.1 model. Pergamon and Elseviers Science, London

Berkes F (2002). Cross-scale institutional linkages: Perspectives from the bottom up. In: Ostrom E, Dietz T, Dolsak N, Stern PC, Stonich S, Weber EU (eds) The Drama of the Commons. National Academy Press, Washington, DC, pp 293-322

Blöschl G, Sivapalan M (1995) Scale issues in hydrological modelling: A review. Hydrological Processes 9: 251-290

Bond G, Showers W, Cheseby M, Lotti R, Almasi P, deMenocal P, Priore P, Cullen H, Hajdas I, Bonani G (1997) A pervasive millennial-scale cycle in north Atlantic Holocene and Glacial climates. Science 278: 1257-1266

Broecker WS (1987) Unpleasant surprises in the greenhouse? Nature 328: 123-126

Chertow M (2001) The IPAT Equation and its variants: changing views of technology and environmental impact. Journal of Industrial Ecology 4: 13-29

Claussen M (1996) On coupling global biome models with climate models. Climate Research 4: 203-221

Claussen M, Gayler V (1997) The greening of the Sahara during the mid-Holocene: results of an interactive atmosphere-biome model. Global Ecology and Biogeography Letters 6: 369-377

Costanza R, Maxwell T (1994) Resolution and predictability: an approach to the scaling problem. Landscape Ecology 9: 47-57

Costanza R, Voinov A, Boumans R, Maxwell T, Villa F, Wainger L, Voinov H (2002) Integrated ecological economic modeling of the Patuxent river watershed, Maryland. Ecological Monographs 72: 203-231

Cox PM, Betts RA, Jones CD, Spall SA, Totterdell IJ (2000) Acceleration of global warming due to carbon-cycle feedbacks in a coupled climate model. Nature 408: 180-184

Crutzen PJ (2002) Geology of mankind: The Anthropocene. Nature 415: 23

Dietz T, Rosa EA (1994) Rethinking the environmental impacts of population, affluence and technology. Human Ecology Review 1: 277-300

Ehrlich P, Holdren J (1971) Impact of Population Growth. Science 171: 1212-1217

Ganopolski A, Kubatzki C, Claussen M, Brovkin V, Petoukhov VK (1998) The influence of vegetation-atmosphere-ocean interaction on climate during the mid-Holocene. Science 280: 1916-1919

Geist HJ, Lambin EF (2002) Proximate causes and underlying driving forces of tropical deforestation. Bioscience 52: 143-150

IPCC (2001) Climate change 2001: synthesis report. Cambridge University Press, Cambridge

Kainuma M, Matsuoka Y, Morita T (2002) Climate policy assessment: Asia-Pacific Integrated Modeling. Springer, Tokyo

Kremen C, Niles JO, Dalton MG, Daily GC, Ehrlich PR, Fay JP, Grewal D (2000) Economic incentives for rain forest conservation across scales. Science 288: 1828-1831

Lambin EF, Turner BL, II, Geist HJ, Agbola SB, Angelsen A, Bruce JW, Coomes O, Dirzo R, Fischer G, Folke C, George PS, Homewood K, Imbernon J, Leemans R, Li X, Moran EF, Mortimore M, Ramakrishnan PS, Richards JF, Skånes H, Steffen WL, Stone GD, Svedin U, Veldkamp TA, Vogel C, Xu J (2001) The causes of land-use and land-cover change: moving beyond the myths. Global Environmental Change Human and Policy Dimensions 11: 261-269

Leemans R (1999) Modelling for species and habitats: new opportunities for problem solving. Science of the Total Environment 240: 51-73

Levin SA (1992) The problem of scale in ecology. Ecology 73: 1943-1967

Lovell C, Madondo A, Moriarty P (2002) The question of scale in integrated natural resource management. Conservation Ecology 5: 25

McCay BJ, Jentoft S (1998) Market or community failure? Critical perspectives on common property research. Human Organization 57: 21-29

McConnell W (2002) Madagascar: Emerald isle or paradise lost? Environment 44: 10-22

Millennium Ecosystem Assessment (2003) People and Ecosystems: A Framework for Assessment and Action. Island Press, Washington DC

Myers N, Kent J (2001) Perverse subsidies. Island Press, Washington DC

Nepstad DC, Verissimo A, Alencar A, Nobre CA, Lima E, Lefebre PA, Schlesinger P, Potter C, Moutinho P, Mendoza E, Cochrane M, Brooks V (1999) Large-scale impoverishment of Amazonian forests by logging and fire. Nature 398: 505-508

O'Neill RV, King AW (1998). Homage to St. Michael: or why are there so many books on scale? In: Peterson DL, Parker VT (eds) Ecological Scale: Theory and Applications. Columbia University Press, New York, NY, pp 3-15

O'Connor J (1988) Capitalism, nature, socialism: a theoretical introduction. Capitalism, Nature, Socialism 1: 11-38

OECD's InterFutures Study Team (1979) Mastering the probable and managing the unpredictable. Report Organisation for Economic Co-operation and Development and International Energy Agency, Paris

Ostrom E, Dietz T, Dolsak N, Stern PC, Stonich S, Weber EU (eds) (2002) The drama of the commons. National Academy Press, Washington D.C.

Palloni A (1994). The relation between population and deforestation: methods for drawing causal inferences from macro and micro studies. In: Lourdes A, Stone MP, Major DC (eds) Population and environment: rethinking the debate. Westview, Boulder, Colorado, pp 125-165

Petchel-Held G, Block A, Cassel-Gintz M, Kropp J, Lüdeke M, Moldehauer O, Reusswig F, Schellnhuber HJ (1999) Syndromes of global change: a qualitative modelling approach to assist global environmental management. Environmental Modelling and Assessment 4: 295-314

Roberts JT, Grimes PE (1997) Carbon intensity and economic development 1962-1971: a brief exploration of the environmental Kuznets curve. World Development 25: 191-198

Rotmans J (1990) IMAGE: An Integrated Model to Assess the Greenhouse Effect. PHD-thesis. Rijksuniversiteit Limburg, Maastricht

Rudel T, Roper J (1996) Regional patterns and historical trends in tropical deforestation, 1976-1990: a qualitative comparative analysis. Ambio 25: 160-166

Schellnhuber H-J (1998). Global change: quantity turns into quality. In: Schellnhuber H-J, Wenzel V (eds) Earth system analysis. Springer, Berlin, pp 12-195

Stafford Smith DM, Reynolds JF (eds) (2002) Integrated assessment and desertification. Dahlem University Press, Berlin

Stern DI (1998) Progress on the environmental Kuznet's Curve? Environment and Development Economics 3: 173-196

Vimeux F, Masson V, Jouzel J, Stievenard M, Petit JR (1999) Glacial-interglacial changes in ocean surface conditions in the southern hemisphere. Nature 398: 410-413

Waggoner PE, Ausubel JH (2002) A framework for sustainability science: A renovated IPAT identity. Proceedings of the National Academy of Sciences 99: 7860-7865

White LJ (1967) The historical roots of our ecological crisis. Science 155: 1203-1207

White R, Engelen G, Uljee I (1997) The use of constrained cellular automata for high-resolution modelling of urban land-use dynamics. Environment and Planning 24: 323-343

York R, Rosa E, Dietz T (2003) Footprints on the Earth: the environmental consequences of modernity. American Sociological Review (in press)

Young O (2002) The institutional dimensions of environmental change. The MIT Press, Cambridge

Index

Printing: Krips bv, Meppel
Binding: Stürtz, Würzburg